殡仪类建筑：
在返璞和升华之间

Funeral
between Nature and Artefact

中文版
（韩语版第355期）

韩国C3出版公社 | 编
王平 朱黛娜 于风军 王莹 | 译

大连理工大学出版社

- 004 感受空间，感受建筑核心 _ Silvio Carta
- 010 Luca Vuerich高山小屋 _ Giovanni Pesamosca Architetto
- 014 骑行天下，自行车客的避风港 _ Rodrigo Cáceres Céspedes
- 020 小木棚屋 _ Rever & Drage
- 024 哈默斯胡斯游客中心 _ Arkitema Architects
- 028 Blåvand地堡博物馆 _ BIG

殡仪类建筑
在返璞和升华之间

- 032 殡仪类建筑：在返璞和升华之间 _ Nelson Mota
- 036 古比奥公墓的扩建 _ Andrea Dragoni Architetto
- 052 韦尔肯拉特丧葬中心 _ Dethier Architecture
- 062 Sant Joan Despí 新殡仪馆 _ Batlle i Roig Arquitectes
- 074 哥印拜陀市的G.K.D. 慈善信托火葬场 _ Mancini Enterprises
- 086 桑塞波尔克罗公墓 _ Zermani Associati Studio di Architettura

社区通道

- 096 社区通道研究 _ Heidi Saarinen
- 100 里斯本广场 _ Balonas & Menano
- 112 圣伊丽莎白东入口亭 _ Davis Brody Bond Architecture
- 118 Weeksville遗产中心 _ Caples Jefferson Architects

住宅
艺术高度

- 130 艺术高度 _ Diego Terna
- 134 图拉住宅 _ Patkau Architects
- 148 独立式住宅 _ Pezo von Ellrichshausen Architects
- 160 悬崖上的住宅 _ Fran Silvestre Arquitectos
- 170 车库房 _ Anonymous Architects
- 178 石屋 _ Olson Kundig Architects

- 190 建筑师索引

004 Sensing Spaces, a Dive into the Core of Architecture _ Silvio Carta

010 Luca Vuerich Mountain Shelter _ Giovanni Pesamosca Architetto

014 Endless Cycling, A Bike-touring Shelter-camp _ Rodrigo Cáceres Céspedes

020 Wooden Sheds _ Rever & Drage

024 Hammershus Visitor Center _ Arkitema Architects

028 Blåvand Bunker Museum _ BIG

Funeral
between Nature and Artefact

032 *Architecture at the Funeral: Between Nature and Artefact _ Nelson Mota*

036 Gubbio Cemetery Extension _ Andrea Dragoni Architetto

052 Welkenraedt Funeral Center _ Dethier Architecture

062 Sant Joan Despí New Funeral Home _ Batlle i Roig Arquitectes

074 Crematorium for G.K.D. Charity Trust in Coimbatore _ Mancini Enterprises

086 Sansepolcro Cemetery _ Zermani Associati Studio di Architettura

Gateway to the Community

096 *Studies of Community Gateways _ Heidi Saarinen*

100 Lisbon Square _ Balonas & Menano

112 St. Elizabeths East Gateway Pavilion _ Davis Brody Bond Architecture

118 Weeksville Heritage Center _ Caples Jefferson Architects

Dwell How
Artificially High

130 *Artificially High _ Diego Terna*

134 Tula House _ Patkau Architects

148 Solo House _ Pezo von Ellrichshausen Architects

160 House on the Cliff _ Fran Silvestre Arquitectos

170 Car Park House _ Anonymous Architects

178 The Pierre House _ Olson Kundig Architects

190 Index

感受空间，感受建筑核心

"人们把建筑作为单纯的塑料造型来看待，犹如看待雕塑和绘画作品，也就是说只看外表，看得肤浅。"[1] 布鲁诺·塞维这一论断是解读凯特·古德温策划举办的"感受空间"建筑展览动机的关键所在。"感受空间：重新想象建筑"这一展览由位于伦敦的英国皇家美术院举办，从2014年一月末开始，于2014年四月初结束。

近些年，大量的展览从各种各样的视角来展现建筑和设计，有的展示建筑对社会的冲击影响，也有的展示建筑作为文化行为的重大意义——建筑可以作为形象，建筑可以表达政治思想。还有许多其他展览聚焦建筑行为和建筑对未来可持续发展可能带来的启示意义，关注当前经济全景或关注技术成就所处的地位。所有这些方面对于关于建筑的争论来说都是非常重要的，但是这些对于像空间、比例或几何结构以及用户对空间的感受这样的建筑核心问题来说只能算是外围次要的方面。

与近些年众多且涉及范围广泛的讨论、辩论以及展览相比，"感受空间"展览似乎瞄准了建筑学科最核心的问题。

本次展会邀请了来自于世界各地的七家建筑事务所来展现建筑形式与人之间的关系。人在建筑作品内感觉如何？建筑材料、材质、光线、形状会怎样影响空间感知？建筑形态和穿越其间的光线能触发人的某种情绪吗？声音、温度、气味和颜色是否也是人们整体的空间体验的一部分？受邀的建筑师致力于通过一系列人们居住的、使用的、甚至是由参观者完成的建筑结构来展示他们的建筑理念。

然而，我们应该注意到，空间感觉取决于教育过程。这一教育过程部分是自然而然发生的，就像是人的身体对外部刺激的自然反应，但是部分还需要训练、关注和阐释。另外，这一"空间意识"因人而异，受年龄因素的影响很大，因为随着年龄的增长，人们的思维方式越来越理性，失去了孩童时对建筑本能的感知和反应。本次展览的介绍性文本（《教师和学生展览介绍手册》，英国皇家美术院出版）就谈到了这一点。手册中提到了英国建筑师科林·圣约翰·威尔逊对空间体验的看法。他说："我们所有的意识都建立在空间体验之上，而这种空间意识不是纯粹单一的，而是充满了我们与生俱来的亲和力所导致的情绪压力。实际上，这是我们所学习的第一门语言，远远早于文字语言。"[2] 在通过理性思维的镜头来看之前，还空间以本来的面目，即其形状及物理特征，这种感受空间的能力对单纯而直接理解建筑来说至关重要。罗伯·格里高利在介绍本次展览时提到了威尔逊的"空间盲"这一概念，即人们缺少按空间本来面目那样感知空间的能力，"一

照片提供：©Chiara Porcu

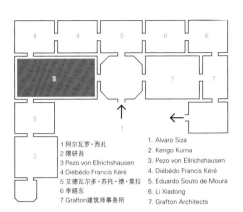

1. 阿尔瓦罗·西扎   1. Alvaro Siza
2. 隈研吾   2. Kengo Kuma
3. Pezo von Ellrichhausen   3. Pezo von Ellrichhausen
4. Diébédo Francis Kéré   4. Diébédo Francis Kéré
5. 艾德瓦尔多·苏托·德·莫拉   5. Eduardo Souto de Moura
6. 李晓东   6. Li Xiaodong
7. Grafton建筑事务所   7. Grafton Architects

在阿尔瓦罗·西扎和艾德瓦尔多·苏托·德·莫拉的项目中，建筑的感知空间被转化为某些典型的元素——西扎选择在英国皇家学院的院子里摆放了柱子和柱顶，而苏托·德·莫拉选择展出模压成型的门框（应用在两种不同的房间内）。
In the Alvaro Siza and Eduardo Souto de Moura's installations, the sensing spaces of architecture are translated into archetypical elements. A column and its capital for Siza in the RA's courtyard and a moulded door case for Souto de Moura (which appears in two different rooms).

照片提供：©Benedict Johnson (courtesy of the Royal Academy of Arts)

种令人困惑也许危险的透明性……一种我们看不到但看得穿的状况"。

所展出的各种建筑设计致力于提高人们的空间感知能力,要求参观者使用他/她的感觉和空间知觉来"感觉"所创建的空间。

隈研吾设计的两个装置被置于进深较深的漆黑房间里,那里唯一的光源是一组嵌在地板上的LED灯。一些竹竿搭建了一个亭子,而LED灯的灯光照在这些竹竿上。黑暗意味着强烈的失落感,因灯光而清晰可见的竹子为人们指引了方向,成为房间里的一个安全点。由于周围一片漆黑,朦胧不清,房间也消失在黑暗中,看不到边缘,看不到墙。黑暗与光明的交织给人以最直接的视觉震撼,而另外两种氛围(扁柏和榻榻米所带来的)则提供了第二种气氛,比黑暗与光明的交织更大范围地弥漫在整个空间中。李晓东的设计是一个迷宫,其中最精彩的部分是一座禅宗花园。低光、镜子,以及人走在鹅卵石上所产生的回声都传达一种迷失方向的感觉。迷宫中通道两侧的墙由淡褐色的木棍包裹,唯一的光源来自于嵌在有机玻璃地板下的LED灯。这一设计的许多方面都让参观者感到不安,促使参观者询问究竟:让所有参观者感到困惑的是整个迷宫没有阴影,因此无法判断周围元素的深度或距离。非洲建筑师Diébédo Francis Kéré的设计故意没有完工,是一件开放的作品,需要参观者的参与来完成。实际上,此设计邀请人们从几个彩色塑料吸管中选取一个,然后插入白色的蜂窝状的基本结构的孔洞里。这一展馆看起来像洞穴,呈毛状,由画廊的两个主房间设计而成。其成品将由参观者完成,每个参观者在特定的时刻所做出的考虑会对展馆做出不断的改变。不像以前的建筑展览,比如上一届威尼斯双年展,展会上展出了许多展馆,阿尔瓦罗·西扎和艾德瓦尔多·苏托·德·莫拉选择了更加成熟而不可或缺的方式。建筑的感知空间被转化为某些典型的元素——西扎选择在英国皇家学院的院子里摆放了柱子和柱顶,苏托·德·莫拉选择展出模压成型的门框。也许最令人印象深刻且空间规模较大的应该算是Pezo von Ellrichshausen和Grafton建筑师事务所设计的作品。Pezo von Ellrichshausen的设计通过使用像坡道、楼梯、阳台或建筑外立面(外立面覆盖住了坡道,也隐藏了展馆其中一部分的内部)这些建筑词汇中的关键元素来展现一种空间体验。参观者可以进入其内,上上下下,参与其内,深入其中。而相比之下,Grafton建筑师事务所的设计用建筑自身的重量和重力来展示其物理方面所唤起的力量。建筑模块和体量几乎全部悬挂在天花板上,天花板上开有天窗,强烈而耀眼的光线倾泻而下,照在各种各样的灰泥表面上。光线增强了墙面和天花板开口处的颜色和色调,把整个空间划分为明暗两块(边上暗,中心亮)。此处可以添加一个旁注:这一结构设计也许会

For me, architecture requires continuity: we have to continue what others have done before us, but using different materials and methods of construction.
– Eduardo Souto de Moura

艾德瓦尔多·苏托·德·莫拉的拱形结构鼓励参观者去考虑Burlington房屋的建筑历史
Eduardo Souto de Moura's arches encourage visitors to consider the architectural history of Burlington House.

According to the ancient Chinese philosopher Lao Zi, what is important is what is contained, not the container.
– Li Xiadong

李晓东的设计是一个迷宫,其中最精彩的部分是一座禅宗花园。低光、镜子,以及走在鹅卵石上所产生的回声都传达一种迷失方向的感觉。
Li Xiaodong's installation is a maze that culminates in a Zen garden where the low light, the mirrors, and the echoes created by people walking on the pebbles convey a feeling of disorientation.

I always start with something small, breaking down materials into particles or fragments that can then be recombined into units of the right scale to provide comfort and intimacy.
– Kengo Kuma

让人们再次意识到，建筑师就是负责掌控着人们的居住空间的实体、上空空间、重量和光线的人。我们可以发现，这一理念体现在本次展览的许多作品中，这些作品展现了建筑实践最基本然而也是最重要的实质。展览室中播放的纪录片更加明确地描述了这一点。Kéré在想建筑物入口处的光线的多少，而西扎的设计意在提醒人们自然是自然，建筑是在空间中建立几何图形的艺术。苏托·德·莫拉认为所有细节都必须为同一个构成体系服务，否则的话就营造不出想要的气氛。Grafton解释说，光线是建筑师拥有的最具表达力的工具之一。因此，建筑不是根据社会、地理或政治情况来呈现的，或只是用模型、图纸或图片来表现自身，而是实实在在且具体地全面呈现建筑这一学科所具有的独特特征。

然而，这一问题需要人们接下来做进一步的思考。虽然展览展出的设计的主要目的似乎是与人们的感官和感知直接互动，但从设计师对光线、材料、色彩以及材质的使用来看，人们不难看出并去欣赏这些设计所具有的象征价值、隐含寓意以及类比手法。另一条把展览中各种不同设计贯穿为一体的、让人不易察觉的思路较清晰地出现在纪录片中。建筑师对建筑的"感受"方式似乎源于童年。孩童时期的一系列感觉总是会伴随我们一生，正如西扎在纪录片中所说，这些感觉在我们进行设计时会再现，有助于我们的设计。这一明确的信息很容易让人联想到威尔逊的关于空间的天真感知的说法：建筑师通过他／她自己的分层情感来构思敏感的空间，反过来，他／她自己的分层情感成为设计师主要的设计工具。对空间的感知是通过被感知的空间形成的。

总而言之，本次展览提醒我们，每一件手工艺品都可以很容易地获得、体现和传递情感、思想和文化价值，但是对于建筑来说，知觉的体验应该是第一位的。另外，本次展览也让我们重新意识到，最难培养的空间阅读能力实际上是我们童年时就已经具有的、与生俱来的能力。

Sensing Spaces, a Dive into the Core of Architecture

"Buildings are judged as if they were sculpture and painting, that is to say, externally and superficially, as purely plastic phenomena".[1] This assertion from Bruno Zevi can be a key to read the motivations of Kate Goodwin, the curator of the exhibition of Sensing Spaces: Architecture Reimagined, organised by the Royal Academy of Arts in London and taking place from the end of January to the beginning of April of 2014. A large number of exhibitions in recent years have presented architecture and design through various lenses, from architecture's impact on society to its significance

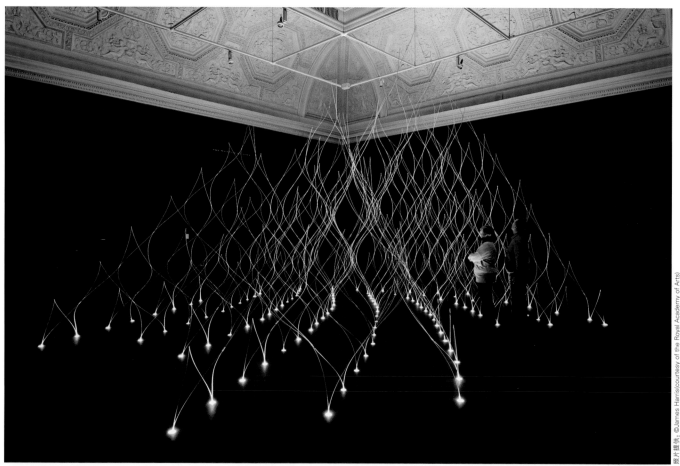

隈研吾设计的两个装置被置于进深较深的漆黑房间里，那里唯一的光源是一组嵌在地板上的LED灯。一些竹竿搭建了一个亭子，而LED灯的灯光照在这些竹竿上。
The two installations by Kengo Kuma are set in deep, dark rooms where the only source of light is an array of LEDs embedded in the floor, pointing at the bamboo sticks that form a pavilion.

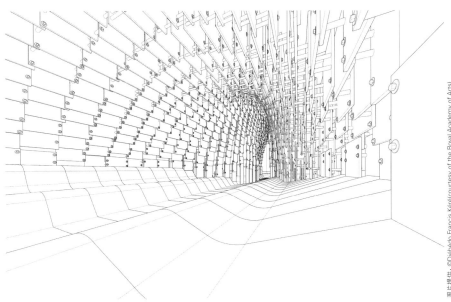

*For me,
architecture is primarily about people,
about asking questions
such as:
Who is the user?
What is going to happen here?
How can I respond to the users' needs?
– Diébédo Francis Kéré*

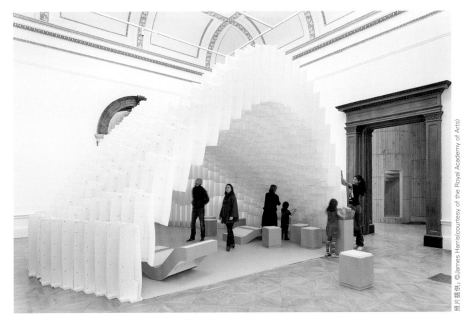

非洲建筑师Diébédo Francis Kéré 的设计故意没有完工，是一件开放的作品，需要参观者的参与来完成。
Diébédo Francis Kéré's installation is purposely incomplete, an open work that requires visitor's intervention for any definition.

as a cultural act – architecture as image and architecture as political expression. Many others have focused on the act of building and possible implications for a sustainable future, with a view toward the current economic panorama or the status of technical achievements. All these aspects are as important as a debate about architecture as such, but all can be considered peripheral to such core questions of architecture as space, proportions or geometry and user perceptions of them.

Against the wide-ranging plethora of discussions, debates and exhibitions of recent years, Sensing Spaces seems to target the very centre of the architectural discipline. Seven architecture practices from around the world have been invited to reflect on the relationship between built forms and people. What does a person feel when inside the architectural work? How do materials, textures, light and shapes affect perceptions of space? Can a built form and the light passing through it trigger a mood? Can sound, temperature, scent and colours also be part of the spatial experience as a whole? The invited architects have sought to demonstrate their points of view through a series of structures to be inhabited, used, or even completed by the visitors.

However, one should consider that the feeling of space depends upon a process of education. Part of this process occurs naturally as a body's reaction to external stimuli, but another part requires training, attention and interpretation. Moreover, the "spatial awareness" varies from person to person and – most significantly – by age, as the gradual acquisition of adult rational thought disables the child's ability to instinctively perceive and respond to architecture. The introductory text of the exhibition (An Introduction to the Exhibition for Teachers and Students, was published by Royal Academy of Arts) touches on this point in its mention of British architect Colin St John Wilson's consideration of the experience of space: *"All of our awareness is grounded in forms of spatial experience, and that spatial awareness is not pure, but charged with emotional stress from our first born affinities. It is in fact the first language we ever learned, long before words."*[2] The ability to perceive space for what it is, for its shapes and physical characteristics, before seeing it through the lens of rational thinking, is fundamental to a pure and direct understanding of architecture. Rob Gregory's introduction to the exhibition refers to Wilson's idea of Spatial Blindness – the incapacity to perceive space for what it is, *"a baffling and perhaps dangerous transparency... a condition that we do not see but see through".*

The various projects exhibited seek to address this point, requiring the visitor to use his/her senses and spatial perceptions to "sense" the spaces created.

The two installations by Kengo Kuma are set in deep, dark rooms where the only source of light is an array of LEDs embedded in the floor, pointing at the bamboo sticks that form a pavilion. While the darkness implies a strong sense of loss, the lit bamboo offers a guide, a safe point in the room, which – owing to the surrounding obscurity – has lost its edges and walls. While the play of darkness and light is the immediate visual perception, the two scents (hinoki and tatami) provide a second, yet more permeating, atmosphere. Li Xiaodong's installation is a maze that culminates in a Zen garden where the low light, the mirrors, and the echoes created by people walking on the pebbles convey a feeling of disorientation. The path offers an unusual experience to visitors with its timber-frame wall clad with hazel sticks, where the only light comes from the LED-lit raised from acrylic floor. Many aspects of this installation prompt visitors to feel uneasy and to ask for reasons: one quandary for all is the total absence of shadows to indicate the depth or distances of surrounding elements in the walkthrough. The installation of the African architect Diébédo Francis Kéré is purposely incomplete, an open work that requires visitor's intervention for any definition. People are in fact invited to take one of several coloured plastic straws and thread it through the holes of the white honeycomb basic structure. The hairy appearance that the grotto-like pavilion assumes – evolving through two main rooms of the Gallery as an enfilade – is hence determined by the visitors and continuously modified according to their considerations at a specific moment. Unlike previous exhibitions, such as the last Venice Biennale, where massive pavilions have been realised, both Alvaro Siza and Eduardo Souto de Moura have chosen a more sophisticated and essential way. The sensing spaces of architecture are translated into archetypical elements – a column and its capital for Siza in the RA's courtyard and a moulded door case for Souto de Moura. Perhaps the most impressive and voluminous proposals are those of Pezo von Ellrichshausen and Grafton Architects. The former proposes an experience of space through key elements of the architectural vocabulary, such as the ramp, the staircase, the balcony, or the facade (which covers the ramp and hides the inside of one part of the pavilion). The visitor partakes of the structure by entering it, circulating up and down, and inhabiting its elements. The latter, by contrast, displays the evocative power of the physical aspect of architecture, with its weight and gravity. Masses and volumes

There is a sense of pleasure in moving from darkness to light or vice versa, because as human beings we're cyclical.
How light reflects and how light is contained are the stuff of architecture.
– Grafton Architects

Grafton建筑师事务所的设计用建筑自身的重量和重力来展示其物理方面所唤起的力量。建筑模块和体量几乎全部悬挂在天花板上，天花板上开有天窗，强烈而耀眼的光线倾泻而下，照在各种各样的灰泥表面上。
Grafton Architects displays the evocative power of the physical aspect of architecture, with its weight and gravity. Masses and volumes are virtually suspended from the ceiling, from which a sharp skylight reflects onto a variety of plastered surfaces.

are virtually suspended from the ceiling, from which a sharp skylight reflects onto a variety of plastered surfaces. The light enhances the colours and tones of both the walls and the voids, differentiating the entire space into dark (on the sides) and lit (at the centre). A side reading can be added here: This structure may recall the fact that the architect is the person in charge of controlling the masses, voids, weights and light of the spaces we all live in. This dawning is supported by a series of suggestions one can find throughout the exhibition concerning the basic yet capital essences of the architectural practice, and is yet more explicitly described in the documentary projected in one of the rooms. Kéré wonders how much light there should be in the entrance of a building, while Siza offers a reminder that nature is nature and architecture is about establishing geometries in space. Souto de Moura holds that details must work together towards the same compositional system, or they don't create the desired atmosphere, and Grafton explains that light is amongst the most expressive tools of the architect. Hence, architecture has not been presented under social, geographical or political circumstances, or as a representation of itself, with models, drawings or pictures, but solidly and concretely, as a holistic sum of the peculiar characteristics of the discipline. However, further and subsequent speculations are left open. Although the main aim of the projects presented in the exhibition seems to interact directly with the senses and perceptions, it is not difficult to see beyond their light, materials, colours and textures to appreciate the symbolic values, the allusions and analogies. Another subtle thread tying the various projects together emerges more clearly in the documentary. The way the "sense" of architecture is envisioned by the architect seems to originate from childhood as the sum of a range of feelings which remain with us in all our lives and which re-emerge – as Siza says in the documentary – to help with our designs. This clear message correlates easily with Wilson's idea of the ingenuous perception of space: the architect conceives a sensing space through his/her own stratified feelings, which are then used as the main items in the designer's toolkit. Sensing spaces are shaped through sensed spaces. To conclude, this exhibition reminds us that each human artefact can easily acquire, embed and communicate feelings, thoughts and cultural values, but for architecture, the sensorial experience should come first. It also reawakens us to the knowledge that the capacity for reading space that is the most difficult to develop is the one we already possessed, innately, as children. Silvio Carta

1. Bruno Zevi, *Architecture as Space*, Horizon Press: New York, 1948, translated 1957.
2. Colin St John Wilson, "The Natural Imagination: An Essay on the Experience of Architecture," *The Architectural Review*, January, 1989, v.185, no.1103, pp. 64~70. Cf. Rob Gregory, *An Introduction to the Exhibition for Teachers and Students*, Royal Academy of Arts, 2014, p.5.

We are not trying to express the structural properties of our buildings.
The emphasis instead is on the proportions of the rooms, their sequence, the way they open – simple things, but when taken together suggests something more complex.
– Pezo von Ellrichshausen

Pezo von Ellrichshausen的设计通过使用像坡道、（螺旋式）楼梯、阳台或建筑外立面（外立面覆盖住了坡道，也隐藏了展馆其中一部分的内部）这些建筑词汇中的关键元素来展现一种空间体验。
Pezo von Ellrichshausen proposes an experience of space through key elements of the architectural vocabulary, such as the ramp, the (spiral) staircase, the balcony, or the facade (which covers the ramp and hides the inside of one part of the pavilion)

Luca Vuerich高山小屋

Giovanni Pesamosca Architetto

Luca Vuerich高山小屋由Diemme Legno-Friulian公司制作和建造。Diemme Legno-Friulian公司专业从事用木质结构和X-lam面板来建造房屋。

这座共有9张床位的小屋,海拔2531m,位于属于朱利安-阿尔卑斯山脉的蒙塔西奥山中的Foronon Buinz群山的山顶,在Ceria Merlone徒步线路一侧。要想感受到惊人的线路、高度和巨石,可以从岔路口Lavinal Bear开始到达此地。高山小屋是为了给登山者、远足者和热爱大山的人们提供一个庇护之所,在此修建也是为了纪念Luke Vuerich:一位登山向导和登山者,被国际登山界看作是一座山峰,在2010年1月Tarvisio附近发生的一次雪崩中遇难,年仅34岁。

宿营小屋由Luca Vuerich家人和高山救援队委托修建。小屋的形状让人们想起了一座小教堂,同时也能承受冬季厚厚的积雪。实际上,小屋的三面几乎完全淹没在雪里,很少见到阳光。阳光明媚的时候,小屋朝南的入口的雪会消融。

小屋占地16m²,木质结构,六根混凝土墩将其抬离岩石地面。小屋利用当地云杉制成的X-lam面板来进行设计和建造,所有构件都在山下制作,用数控机床切割成合适尺寸,然后用直升机运到山顶场地,就地按照技术图纸进行组装。

设计完成之后,建筑材料开始进行生产加工,共需要30块X-lam面板、3个桁架,还有落叶松木材制成的基座。建筑材料生产加工完成后,就到了规划阶段,即组织墙体和其他材料(全都标号且准备组装)的运输以及协调安排所需的建筑人员:Diemmelegno的工人和技术人员、高山救援志愿者和Luke的朋友们,一共有12个人愿为宿营小屋的建造助一臂之力。直升机往返18次把建筑材料运到蒙塔西奥山上,建筑人员等在山顶,条件艰苦,工作空间狭小,先是把材料从空中盘旋的直升机上卸下,然后安安全全地把小屋组装好。所有这一切都要在一天之内完成。施工队在此要住一个晚上,第二天,他们要完成小屋最后的修整和涂漆工作。从小屋落成的那天起,无论寒冬还是盛夏,它就已经成为登山爱好者的目的地,成为一个安全之所、保护之所,观赏令人窒息的山中美景,看山羊在岩石间跳跃,一切淹没在大自然的沉寂之中,似乎必不可少。

Luca Vuerich Mountain Shelter

Manufactured and built by Diemme Legno - Friulian Company that manufactures and constructs buildings with wooden structure and panel X-lam.

The structure with 9 beds is located at an altitude of 2531 meters in the Julian Alps, on the crest of the Foronon Buinz Group of Montasio along the trail equipped Ceria – Merlone which starts from fork Lavinal Bear for a breathtaking route, altitude the giant stone. It was designed as a shelter for mountaineers and hikers, for lovers of the mountain, and built to remember Luke Vuerich: a mountain guide, climber considered a peak in the international scene who died in January 2010 at age of 34 because of avalanche near Tarvisio.

The camp was built and commissioned by the family of Luke Vuerich along with the Mountain Rescue section. It not only remembers to form a chapel, but is also designed to support heavy snow loads that can be stored during the winter. In

项目名称:Bivacco Luca Vuerich
地点:Udine, Italy
建筑师:Giovanni Pesamosca Architetto
结构工程师:Ingegnere Valentina Bertolutti, Diemmelegno snc, I.D.I.R. srl, Lattoneria De Cillia, Baron Maurizio
工厂:Studio T.E.A.
顾问:Ingegnere Marco Pesamosca, Geometra Vittorio Di Marco
主管:Geometra Roberto Palmieri
甲方:Luciano Vuerich
表面积:15m²
体积:30m³
设计时间:2011.12~2012.4
施工时间:2012.7
竣工时间:2012.8
摄影师:©Flavio Pesamosca(courtesy of the architect)

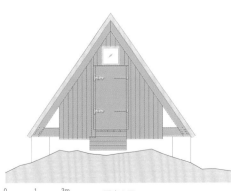

0 1 2m

西南立面
south-west elevation

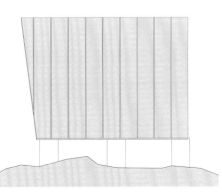

东南立面
south-east elevation

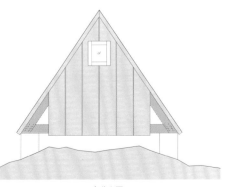

东北立面
north-east elevation

fact, it is almost completely submerged in the three sides. So they are less exposed to the sun, whereas it reveals a south-facing access through the work day of the sun.

The property with an area of 16 square meters is made of wood, and is raised from the rocks making 6 rest on concrete piers. Designed and built with the building system panel X-lam with local spruce, the elements that compose it have been manufactured and cut to size with a CNC machine on the ground, transported to the site by helicopter and assembled several trips on the spot according to the technical papers.

After completion of the design and production of the structure - made up of 30 panels X-lam, 3 trusses and main base in larch wood – began the organization phase to allow the transportation portion of the walls and several pieces (all numbered and ready for assembly) as well as coordinate the work force: workers and technicians of Diemmelegno, mountain rescue volunteers and friends of Luke, a total of 12 men are all ready to "lend a hand" to build this camp. The material, once arrived on the plateau of Montasio was transported by helicopter to share with 18 journeys. Waiting for it on the top, those who work in difficult conditions and with little space first dropped the packs from the aircraft suspended in flight and then mounted the property in total safety, all within a single day. After a night spent in the bivouac, the next day they made the works of finishing and coating. From the day of the inauguration, it has become a destination for mountain lovers both during the harsh winter and summer, as a safe place and protection with a breathtaking mountain view and goats among rocks, immersed in the silence of nature, as if it is essential.

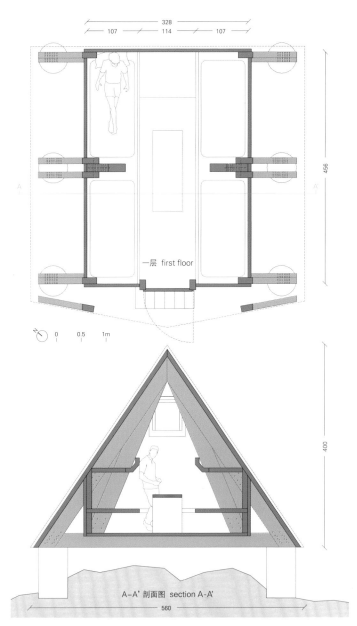

一层 first floor

A-A' 剖面图 section A-A'

旅行者的小天地 Travellers' Small

骑行天下，自行车客的避风港 _Rodrigo Cáceres Céspedes

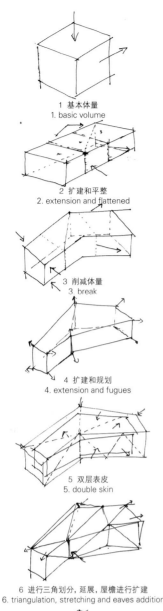

1. 基本体量
1. basic volume

2. 扩建和平整
2. extension and flattened

3. 削减体量
3. break

4. 扩建和规划
4. extension and fugues

5. 双层表皮
5. double skin

6. 进行三角划分、延展、屋檐进行扩建
6. triangulation, stretching and eaves addition

7. 进气口、斜坡以及风幕
7. air intakes, slopes and wind shadows

8. 最后的造型和体量
8. final shape and volume

这一设计是塔尔卡大学建筑学部毕业典礼之前的一个个人委托的项目。设计关键词包括课外活动、一处区域、可移动的结构、以及一种服务。所提供的服务兴起于在旅游业和体育的灰色地带中发展职业生涯的良机。设计师把下列因素融为一体就是本设计所形成就的作品：一群骑自行车的人，一天中不必返回起点的连续不断的骑行；在多样而广袤的大地增建一个结构，为骑行者提供在一天中或几天中观赏不同景色的机会；考虑提供一个装置（建筑），既可作为出发点和终点线，也可作为骑行者恢复休整和与其他骑行者会合的地方；在建筑物内配备一些简单但实用的物品作为后勤保障，使自行车爱好者和他们的朋友们每年秋天、春天和夏天都可以回到此地，进行运动锻炼的同时也可以探索这一区域。

第一步是清楚标记这一区域。目的有二：1）向骑行者和游客标出各骑行线路；2）强调这一项体育运动说明了人们开始选择新的方式来欣赏伊甸园般令人愉悦的大自然。下一步与本设计的移动性相关。设计的装置要比帐篷更加舒适，但是同时又要易于安装和拆卸，只要小部分人就可以安装或拆卸，在任何情况下都不需要自行车手的参与，因为他们的主要任务就是骑行和欣赏山谷景色。下一步围绕如何最大限度地将装置置于自然之中，被大自然所包围。一方面，这一设计原型所在的路线各不相同，在一次探险的同一路线上不能重复设置。在下一站，这一设计原型需要承担不同的角色，比如，作为一个观景点、一个补给站、或休息处，这样，骑行者可以有机会饱览风光景色，同时场地也为他们提供了一处尽情享受整个体验的地方。以这种方式，建造者达到了最终目的，即把自行车手带到一个远离像自行车赛道或奥林匹克自行车馆这样受限制的、场地以外的地方。最后一步是创建一个新系统。许多原型可以帮助创建设计中必要的功能特点。这些功能特点可以满足整个地区众多自行车手的一次性使用，充分体现了自行车活动是本地区的一项重要的赛事和活动。

设计的小营地内还有靠踩踏板摩擦发电的发电机，当发电机连接到交流发电机和电池时，就可以为一组LED灯提供电能。水储存在水箱里，靠重力到达饮水机里。太阳能18-lt淋浴器可以为热水浴提供所需的水量，其设计安装是整体结构的一部分。污水和垃圾将进入不同的容纳箱，被分别安置在合适的位置。

Endless Cycling, A Bike-touring Shelter-camp

The project is a personal commitment prior to the graduation ceremony of the Architecture Faculty at the University of Talca which includes extracurricular activities, an area, a movable construction, and a service which has emerged from an opportunity to develop a professional career in a grey area existed in the tourism industry and sports. Here lies the fusion that allows to build the experience: take a group of cyclists and their ability to move

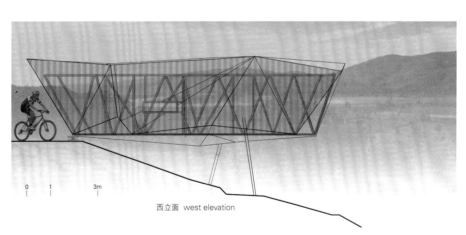

西立面 west elevation

南立面 south elevation

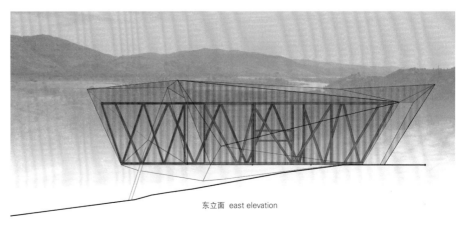

东立面 east elevation

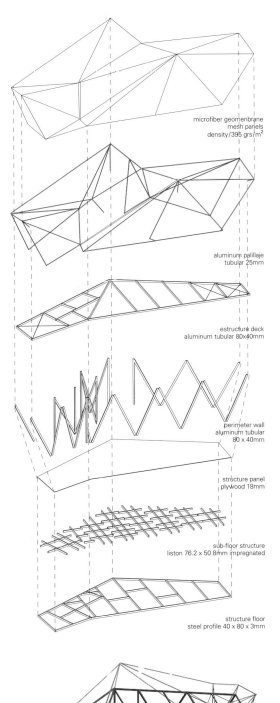

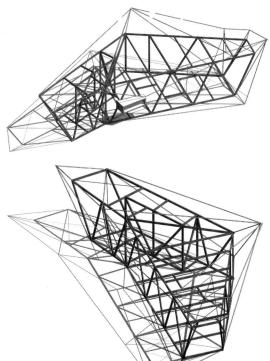
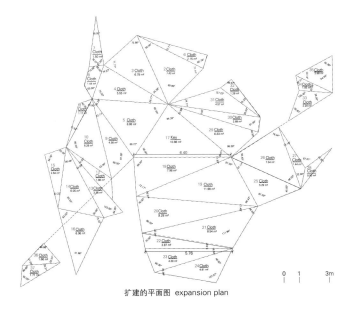

扩建的平面图 expansion plan

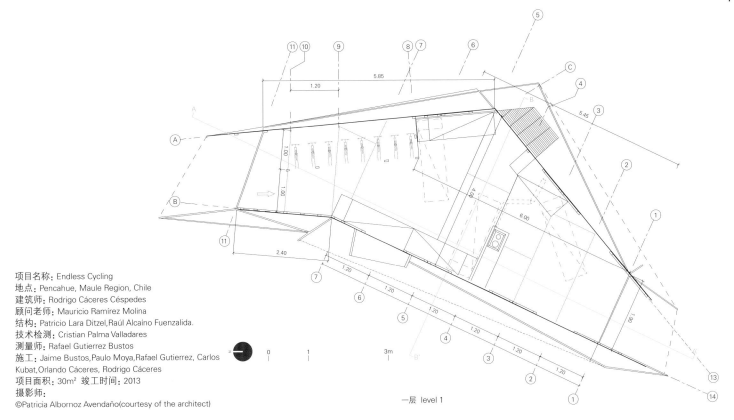

一层 level 1

项目名称：Endless Cycling
地点：Pencahue, Maule Region, Chile
建筑师：Rodrigo Cáceres Céspedes
顾问老师：Mauricio Ramírez Molina
结构：Patricio Lara Ditzel, Raúl Alcaíno Fuenzalida.
技术检测：Cristian Palma Valladares
测量师：Rafael Gutierrez Bustos
施工：Jaime Bustos, Paulo Moya, Rafael Gutierrez, Carlos Kubat, Orlando Cáceres, Rodrigo Cáceres
项目面积：30m² 竣工时间：2013
摄影师：
©Patricia Albornoz Avendaño(courtesy of the architect)

uninterruptedly without the need to come back to the starting point on the same day; add such a diverse and vast expanse of land which could give the cyclists the opportunity to observe the different landscapes on the same or more days; consider a device (the construction) that serves not only as the starting point and finish line, but also as a place to recover and meet the other cyclists; and, merge this with simple but effective logistics that will make the cyclists and their friends come back every Fall, Spring and Summer in order to explore the area while doing sports at the same time.

The first approach is to mark the area for two main reasons: 1) show the trails to the cyclists and tourists, and 2) show that this alternative sport is showing that people are starting to choose new ways of enjoying Eden's blissful copy. Approach Number 2 has to do with mobility. The device has to be much more comfortable than a tent, but, at the same time it needs to be easy to be set up and dismantled by a small group of people, which under no circumstances will involve the cyclists since their sole role is to pedal and enjoy the valley. The next approach revolves about maximizing the fact of being surrounded by nature. On the one hand, the routes on which this prototype is to be taken need to be different; the device must not be taken through the same route twice in one expedition, and during the next stop, the prototype has to adopt several roles, such as a viewpoint, supply stop, or aplale for relaxation that can give the cyclists the opportunity to enjoy the landscapes and provide them with a place where they can wallow in the whole experience. In this way the final goal, which is to take the cyclist to a place other than the restricted scenarios like cycle tracks or velodromes, is achieved. The last approach is related to the creation of a new system. Many prototypes can help to create the necessary features for a design that will allow a multiple number of cyclists to go all over the area in use in just one go, which maximizes its presence as an important event and activity in the area.

A power generator that works by pedaling friction has been also added, and which, when connected to an alternator and a battery, sends energy to a set of LEDs. Water is stored in a tank and is taken to the dispensers using gravity. A solar, 18-lt shower will provide the needed amount for a hot-water bath, and is part of the structure itself. Sullage and waste will go into different tanks so they are placed in the right locations.

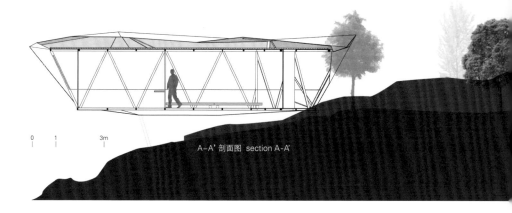

A-A' 剖面图 section A-A'

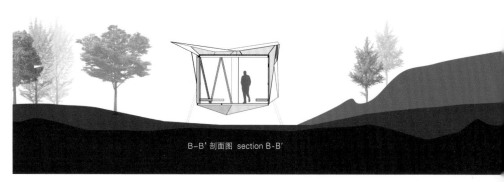

B-B' 剖面图 section B-B'

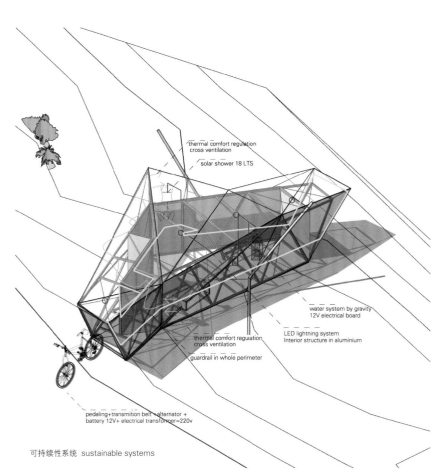

可持续性系统 sustainable systems

小木棚屋 _Rever & Drage Architects

这座小巧但功能多样的建筑是按照客户的需要来设计和建造的，一方面是在夏季房子外面的露天广场上有一处可以遮风挡雨的地方，另外既可以用来存放工具，又可以在特别的时候用来在星星下小睡。一座小小的建筑，但设计非常复杂，既要包含双重功能的元素又要体现建筑的模糊性。

项目所在地位于挪威最西北部的海边，这里有的时候天气条件非常恶劣，且总是变化无常，每天都会有飘来的海水。

最终，这座建筑看起来既新颖又古朴。建筑的主要形式使用了抽象的建筑语言，也没有装檐口，是典型的现代风格，而建筑的外表面是典型的老式学院派风格，经过包装的木板材就像使用了当地小木船的传统的防水处理方式。树皮不仅使木材产生更深的视觉深度，也使这座建筑在夕阳的余辉中看起来更加迷人。

当房门紧闭时，这座建筑让人感觉就像是一位年长而谨慎的圣女，为冬日的暴风雨做好了准备；而在仲夏的夜晚，房门敞开，有了修饰装扮，她就如一朵盛开的花朵。整体看起来又像是巨石阵，高大厚重，似乎是从成百上千英里以外运到这里的。

如果有太阳，但人们还觉得北风有点儿清冷（这一地区典型的天气情况）的话，可以把较小的那个棚屋的门拉开，就形成了一个围合起来的小广场。同时，因为小棚屋的后墙安装的都是玻璃，因此仍能欣赏到海景。如果天气暖和，但空中飘些细雨，就可以电动打开主建筑物的上层屋顶，同时可以看到上层屋顶下面的天窗。从防水方面来说，这个玻璃屋顶是建

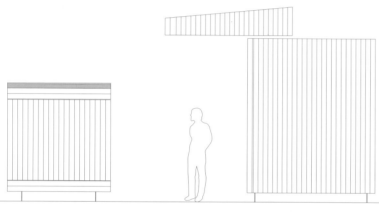

南立面_设备棚
south elevation_equipment shed

南立面_附属建筑物
south elevation_annex

筑的主屋顶，引导水流流到建筑的后面而不是流到广场上。而上面的木屋顶的倾斜度恰恰相反，以面对较强的西风，同时在冬天可以承受住积雪。

设计中屋顶的前后滑动让人或多或少地想到了莱昂纳多·达·芬奇的作品：有轮子、钢丝电线、滑动的梁和衡重体。在这一制造问题和解决问题的过程中，这座建筑让懂得工程技术的路人报以兴趣盎然的微笑，解决了最初的问题，也令客户感到满意。

Wooden Sheds _ Rever & Drage

This small but multifunctional building was designed and constructed, both as an answer to the client's need for a wind-and-rain shelter at their outdoor summer house-piazza, and as a combined tool-shed and special-occasion-sleep-under-the-stars-facility. It's a complex program for a modest building, making way for double-functional elements and architectural ambiguity.

The site is at the utmost north-western-

coast of Norway, presenting itself with some harsh and always changing weather conditions including a daily spray of salt water.

Finally the building turned out to be both new and old. The main forms, in their abstract expression and lack of cornice, are typical modern looking, while the exterior surface is typical old-school with the wood panels coated just like the traditional waterproofing for local wooden boats. The tare, whilst bringing out the visual depth of the wood, also makes the building quite charming in the low evening sun.

The building in its closed position gives

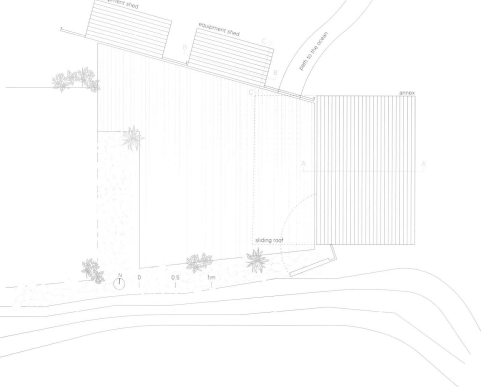

项目名称：Hustadvika Tools
地点：Hustadvika, Norway
建筑师：Rever & Drage Architects
项目团队：Tom Auger, Martin Beverfjord, Eirik Lilledrange
面积：15m²
竣工时间：2013

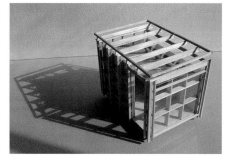
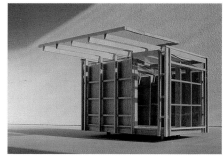

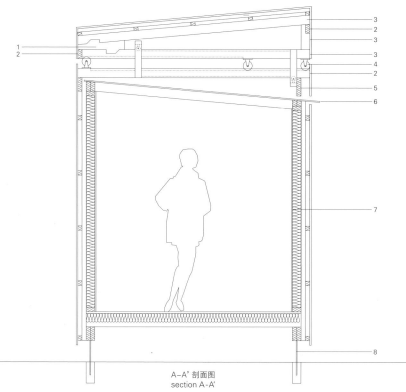

A-A' 剖面图
section A-A'

1. timber hook joint
2. 48 x 98mm
3. 98 x 98mm
4. wheels, height 128mm
5. 48 x 198
6. lexan roof
7. OSB board
8. steel post base
9. timber cladding 17 x 100mm
10. wheels
11. timber cladding 20 x 100mm
12. 48 x 48mm
13. lexan, 10mm
14. gutter
15. OSB board, 12mm
16. shelf
17. OSB board, 18mm
18. 48 x 73mm
19. 36 x 48mm
20. 48 x 148mm

somehow the impression of an old prudent virgin preparing herself for the winter storms, whilst in its open position it is a decorated shed blooming in the midsummer night. All over the final result is also a Stonehenge-like place with its high and heavy features transported there from hundreds of miles away.

If the sun is out, but the northern wind is a bit chilly (which is a typical condition in this area), sliding out the doors from the smaller sheds will form a continous embracement of the small piazza. At the same time the back walls of the sheds are made of glass, such that the ocean view is maintained. If the weather is warm, but there is some rain in the air, the upper roof of the main building can be slid out by an electrical engine, simultaneously uncovering a skylight inside. This glass roof is the main-roof of the building in terms of waterproofing, leading water away from the piazza to the back of the building, whilst the wooden roof on top is tilted the opposite way, to face the stronger western winds and also take the snow burden during winter.

Making the roof slide back and forth gave the project a (tiny) hint of Leonardo da Vinci's activity, with its wheels, wires, sliding-beams and counter-weights. In this problem-making, as much as problem-solving process, the building generates interested smiles from engineer-hearted passers-by, as well as solves the original program and satisfies the clients.

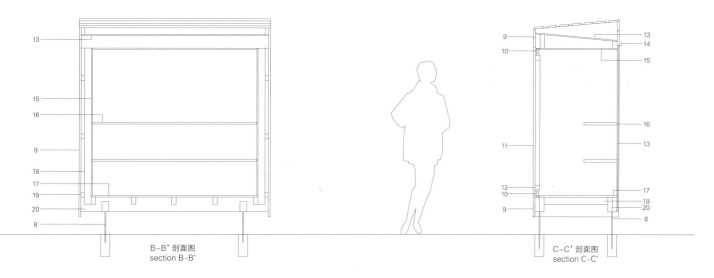

B-B' 剖面图
section B-B'

C-C' 剖面图
section C-C'

文化风景线 Culturescape

哈默斯胡斯游客中心 _Arkitema Architects

令人梦寐以求的竞赛项目

新游客中心将耗资4500万丹麦克朗,包括展览区、咖啡馆和用于教育目的的房间,每年将为大约500 000名游客提供服务。共有53家建筑公司想要参与设计竞赛,后来挑选出六个团队,最终Arkitema建筑师事务所成为最后的赢家。

融于风景中的游客中心

"我们的设计从属于大自然,使哈默斯胡斯成为主角。与此同时,我们充分利用项目位置,无论从游客中心的哪一个角度,都使哈默斯胡斯成为焦点。新游客中心将从哈默斯胡斯城堡遗址宏伟壮观的景色中消失,成为风景的一部分,成为自然的一部分,是有价值的增建建筑。"Arkitema建筑师事务所的合伙人和建筑师Poul Schülein如是说。

更具体地说,游客中心的屋顶与地势融为一体,并一直延伸至这一地区的公共人行步道中。这样,屋顶就成为观景平台,可以观赏到遗址叹为观止的景色。屋顶也成为公共空间,就像以往一样,无论是夏天还是冬季,无论是白天还是夜晚,游客都可到此——景观中的一个热情的文化印记。

重视环境的建筑材料

"我们设计了一座很简单的房子,形状细长,精巧地融于地势之中,其屋顶设计非常重要。同时,建造房子所使用的材料也很简单,只使用木质材料和混凝土,并且都取自周围环境。混凝土中的岩石颗粒就是来自于为修建游客中心开山劈石而劈下来的石块。我们的设计无论是形式还是使用的建材都充分地考虑和维护项目所在地的历史背景。"Poul Schülein总结道。

游客中心的实用功能部分都被设计在靠近峭壁的一侧,而另一侧设计了大型的窗户,面向令人叹为观止的哈默斯胡斯城堡遗址、优美的风光和大海。

其他五支参与竞争的团队是Vandkunsten、KHR、Julien De Smedt、Cubo和Entasis。

游客中心将于2016年春天对游客开放。

Hammershus Visitor Center

A Much-Coveted Competition
The new visitor center will cost 45 million DKK. It includes an exhibition, a cafe and educational rooms and will serve approximately 500,000 annual visitors. A total of 53 architectural firms wanted to be in the architectural competition, six teams were then picked out and Arkitema was elected the final winner.

A Visitor Center Integrated in the Landscape
"Our proposal subordinates to the nature

and allows Hammershus to be the main attraction. At the same time we exploit the location to create full focus on Hammershus from the visitor center. From the ruin the new visitor center will appear as a contribution to the landscape – an integrated part of nature. A worthy addition." says partner and architect at Arkitema, Poul Schülein.

More specifically the roof of the building is incorporated in the terrain, where it continues the public footpaths in the area. In that way it establishes a scenic platform with a stunning view to the ruin. The roof establishes a public space, which gives something back to the place by being available to the visitors both summer and winter, day and night – a welcoming cultural imprint in the landscape.

Materials that emphasize the environment

"We have designed a simple house with an elongated shape that is carefully sited in the terrain, and where the roof is an important factor. At the same time the materials used in the house are taken from the surroundings – simple wooden materials and concrete. The concrete consists of rock-granulate from the cliff that we dig out to make room for the building. We have designed a building that in its form and materials upholds and stages its historical context." Poul Schülein concludes.

Towards the cliff, the practical functions in the house are placed and to the opposite side the house opens up with large windows facing the stunning view of Hammershus, the landscape and the sea. The other teams in the competition were Vandkunsten, KHR, Julien De Smedt, Cubo and Entasis.

The visitor center will be opened in the spring of 2016.

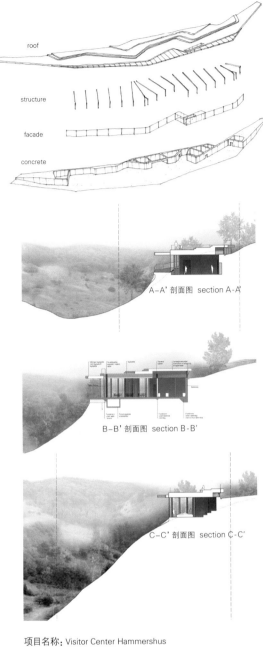
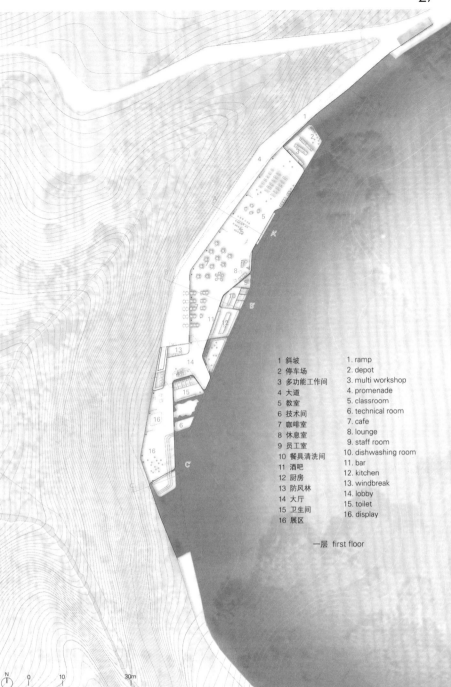

1 斜坡	1. ramp
2 停车场	2. depot
3 多功能工作间	3. multi workshop
4 大道	4. promenade
5 教室	5. classroom
6 技术间	6. technical room
7 咖啡室	7. cafe
8 休息室	8. lounge
9 员工室	9. staff room
10 餐具清洗间	10. dishwashing room
11 酒吧	11. bar
12 厨房	12. kitchen
13 防风林	13. windbreak
14 大厅	14. lobby
15 卫生间	15. toilet
16 展区	16. display

一层 first floor

项目名称：Visitor Center Hammershus
地点：Hammershus, Bornholm, Denmark
建筑师：Arkitema Architects
景观建筑师：Arkitema Architects
工程师：Buro Happold
其他：Christopher Harlang
甲方：Danish Forest and Nature Agency
用地面积：1,355m² 设计时间：2013

文化风景线 Culturescape

Blåvand地堡博物馆

Blåvand博物馆中心由4个独立的部分组成，分别是地堡博物馆、琥珀博物馆、历史博物馆和一个专题展览画廊，融于嵌入沙丘之内的展览景观之中。与坚固厚重的地堡掩体形成对照，Blåvand博物馆看起来就是景观中经过精确切割的若干部分的交汇点，成为景观中的一座建筑，从而将沙丘掩藏起来。

堡垒中的枪支被修复成一个幽灵，或者可以说是这一战争机器的本来面目的反射。大炮采用玻璃和钢框架来重新建造，是一件线框勾勒出的透明的人工制品，位于冰冷的地堡之上，成为地堡的天窗，游客可以在此充分感受在"大西洋壁垒"防御工事中枪支所占据的重要战略位置。因此，这座新建筑既是对现有堡垒建筑的批判，也体现了对其的尊重。与堡垒建筑形成对照，新建筑轻盈，舒适，宽敞而不厚重，透明而不沉重。

玻璃大炮和隐形的博物馆同时敏感地添加于现有景观之中，融于自然之中，只有走进，走进沙丘，走进地堡，游客才能揭开其面纱。

Blåvand Bunker Museum _ BIG

Museum Center Blåvand integrates 4 independent institutions – a bunker museum, an amber museum, a histolarium and a special exhibition gallery – in an exhibition landscape embedded in the dunes. As the antithesis to the heavy volume of the bunker, the museum appears as the intersection between a series of precise cuts in the landscape. It's a block in the

项目名称：Blåvand Bunker Museum
地点：Varde, Denmark
建筑师：BIG
主要合伙人：Bjarke Ingels, Jakob Lange, David Zahle, Andreas Klok Pedersen
项目负责人：Brian Yang
项目团队：Michael Schønemann, Alina Tamosiunaite, Katarzyna Siedlecka, Ryohei Koike
甲方：The Museum of Varde City and Vicinity
用地面积：2,500m² 竣工时间：2012

4 museums
Museum center Blåvand contains 3 museums and 1 special exhibition gallery.

whole
At the same time the complex is seen as a larger entity with the possibility for continuous circulation and communication between the various museums.

autonomous
Each museum serves as an independent unit with requirements for the ability to change exhibitions, host special events and have their own opening hours.

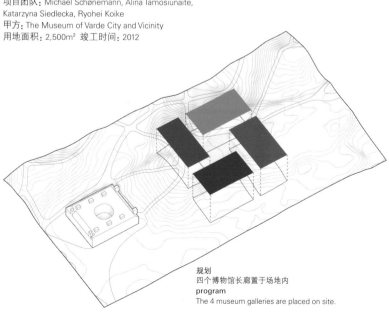

规划
四个博物馆长廊置于场地内
program
The 4 museum galleries are placed on site.

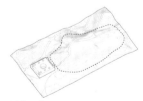

existing site
The bunker and crane embankment to its north constitute the site for the new museum.

integration into the landscape
4 simple cuts are made into the topography, creating a central courtyard for the museum. The 4 paths connect back into the existing netweork of trails in the dune landscape.

rotation of foyer
with a simple rotation of the foyer area

daylight
4 skylights at each corner bring daylight into the center.

upper level: arrival and cafe
Entrance to the museum is via the central courtyard and into the cafe.

cafe

lower level: exhibition & support functions
A common foyer is located in the center of the galleries on the lower level. Operational support functions such as toilets, cloakroom, and exhibition storage are placed between each gallery space.

support functions

exhibition access
Create an opportunity for access to the 4 museum galleries.

connection to the Tirpitz Bunker
An underground tunnel connects back to the old Tirpitz Bunker.

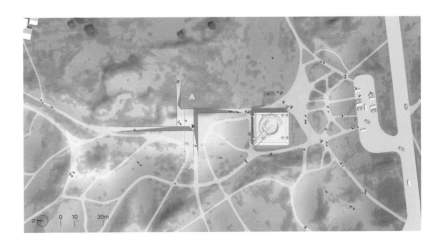

- arrival
- lobby
- special exhibition

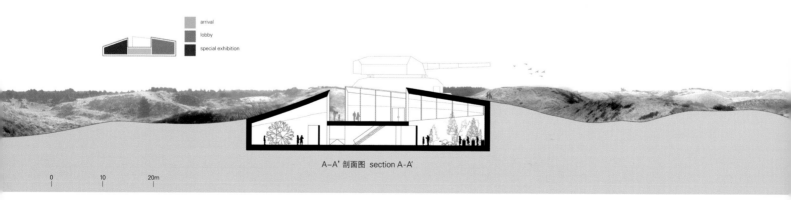

A-A' 剖面图 section A-A'

0 10 20m

landscape – and a corresponding absence of the dune.

The bunker's gun is restored as a ghost or a reflection of the war machine it was meant to be. As a crystal artifact in "wireframe" – the cannon is re-created in glass and steel framing – it is a skylight over the raw setting of the bunker where visitors can experience the strategic position the gun would have played in the fortifications of the Atlantic Wall. The new architecture is at once critical and respectful to the existing bunkers. As an antithesis – vacuum rather than volume – transparency rather than gravity – it represents the new architecture of a light and easy antithesis of bunker architecture. Simultaneously, both the glass cannon and invisible museum will add sensitively to the existing landscape and nature which only on a closer inspection – a walk in the dunes, or a visit to the bunker – unfolds for visitors.

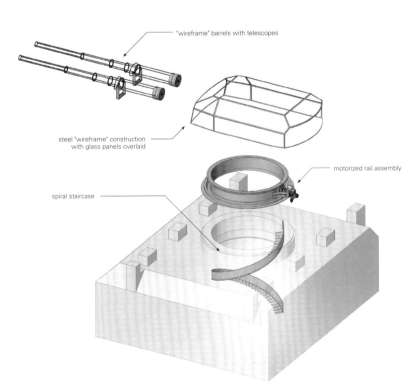

殡仪类建筑：在返璞和升华之间

Funeral
between Nature and Artefact

在自然与人造之间寻求一种平衡，这在建筑设计中极为常见。自远古以来，建筑与景观之间就存在着一种辩证关系，建成环境的转变挑战着自然现象的平衡。殡仪类建筑便是其中之一。殡仪类建筑的设计者必须理解死亡这种自然现象，并且要使它们与自己的反面相协调。实际上，丧葬是一种仪式，它存在于生与死、记忆与遗忘、愤恨与平静、团体与隔离之间的阈限空间中。在那一时刻，矛盾的个体间会有激烈的碰撞，从而引发了际遇不同的各方面的争论。

有证据表明，一些举世闻名的考古发现正是古代的坟墓所在地或是丧葬遗迹。它们都是天人合一的佐证。带景观的建筑是死亡权术的固有因素。而后者主要是受到文化的影响，也就是说它并不是适用于一切的普遍真理。因此，殡仪类建筑能够表现建筑是如何与时间和空间相互调和的。此外，墓地、火葬场以及其他一些与殡仪服务有关的地方都有着很明显的特征。并且，它们所承担的社会服务已经超越了公共或是私人礼仪设施的范围。

因此，殡仪类建筑不得不在人类有限的生命时间与无限的永恒之间进行调和。这就需要借助空间装置进行创造，从而迎合人类世界观中的宇宙论。建造这些协调空间也正强调了边界划分的重要性，比如说公共领域与私人领域、建筑与自然，或者是个人与社会。

Producing a balance between nature and artefact is one of the most recurrent facets in architectural design. Since immemorial times architecture and landscape developed a dialectical relation where the transformation of the built environment challenged the balance of natural phenomena. The architecture of funerals is one of these events where the designer is called to make sense of a natural phenomenon, death, and to reconcile opposite aspects. In effect, a funeral is a ritual that dwells in a liminal space between life and death, memory and oblivion, resentment and quietness, communion and seclusion. It is a moment that calls for an intense negotiation of conflictive spirits and brings forth a confrontation of opposite experiences.

Arguably some of the most well known archaeological findings are ancient burial spaces or funerary monuments. They testify to the fact that reconciling men with nature, and architecture with the landscape are immanent components of the politics of death. The latter is, nevertheless, a phenomenon chiefly influenced by culture, which suggests that no universal truth is applicable to it. The architecture of funerals is thus an eloquent repository on how architectural operations negotiate time and place. Furthermore, cemeteries, crematoriums and other places where funeral services are performed often carry strong symbolism and perform a social role that goes beyond its mere function as a public or private amenity.

Hence, the architecture of funerals has to negotiate the transience of our earthly existence with the timelessness of the sublime and the eternity. It calls for the creation of a spatial device that has to cater for a specific cosmological view of the world. Giving shape to these spaces of negotiation stresses the importance of the boundaries dividing such aspects as the public and private realm, the tectonics of construction and Nature, or the individual and the community.

古比奥公墓的扩建_Gubbio Cemetery Extension/Andrea Dragoni Architetto
韦尔肯拉特丧葬中心_Welkenraedt Funeral Center/Dethier Architecture
Sant Joan Despí 新殡仪馆_Sant Joan Despí New Funeral Home/Batlle i Roig Arquitectes
哥印拜陀市的G.K.D.慈善信托火葬场_Crematorium for G.K.D. Charity Trust in Coimbatore/Mancini Enterprises
桑塞波尔克罗公墓_Sansepolcro Cemetery/Zermani Associati Studio di Architettura
殡仪类建筑：在返璞和升华之间_Architecture at the Funeral: Between Nature and Artefact/Nelson Mota

建筑与景观

由瑞典建筑师Erik Gunnar Asplund及Sigurd Lewerentz所设计的森林墓地位于斯得哥尔摩郊外，该设计始于1915年前后，于20世纪30年代末完成，堪称北欧现代主义建筑的代表作。设计师做到了返璞与升华之间的完美平衡，从而达到了不同寻常的效果。建筑与景观成为了单一的实体，模糊了传统意义上的区别。这个建筑体现了地形学、建筑外形及材料运用的一种独特感觉。可以确定的是，这些建筑采用的都是一些惯常使用的方法，我们在其他的北欧建筑师的作品中也可以看到，如阿尔瓦·阿尔托、约翰·伍重及斯维勒·费恩，略举数例。伍德兰公墓其实在1994年就已经被联合国教科文组织列为世界遗产，足以见得它的独特性。

说说离我们更近一些的时代吧，由西班牙建筑师恩里克·米拉列斯及加尔尼·皮诺斯建造的伊瓜拉达墓园（1984—1994）展示了建筑与景观的另一段完美对话。这座建筑不像之前提到的瑞典建筑那样能够勾起人们的回忆，但是建筑材料的使用是定义伊瓜拉达墓园整个氛围的关键因素。米拉列斯和皮诺斯使用水泥及考顿钢来刻意强调大自然的粗放与人造材料间的对比，这些材料自然而然地成为了新景观的组成元素之一。这两例建筑都透露出建筑师想要让阴阳两界相互融合的明确意图。

在这些公墓中，记忆与回忆唤醒了今天的人们，同时也勾画了未来。此外，在斯得哥尔摩及伊瓜拉达，建筑的细枝末节都体现了与自然尖锐的碰撞，这种设计也恰恰是一种世俗特征。自然界的灵魂是瞬息的，这也正好与人类每天的不断建造达成了一种平衡。在我们之前所提到的建筑项目中，有种调和感，比如建筑与景观，神圣与世俗，这些在不同国家的建筑中都是不可或缺的特点，比如在西班牙、意大利、比利时、印度及日本的建筑。

Architecture and Landscape

The Woodland Cemetery designed from the mid-1910s until the end of the 1930s by the Swedish architects Erik Gunnar Asplund and Sigurd Lewerentz for a location in the outskirts of Stockholm, is arguably one of the masterpieces of Nordic Modernism. The architects' masterful balance between nature and the artefact produced an unusual outcome: architecture and landscape became a single entity, blurring the traditional distinctions between them. This operation was accomplished through a particular sensitivity to the manipulation of topography, form, and to the use of materials. To be sure, these are indeed customary tokens in the architectural approach of other Nordic architects, such as Alvar Aalto, Jorn Utzon, and Sverre Fehn, to name but a few. The Woodland cemetery is actually listed in Unesco's World heritage sites since 1994, which testifies to its unique character.

More recently, the Igualada Cemetery (1984~1994) designed by the Spanish architects Enric Miralles and Carme Pinos shows another instance of this delicate dialogue between architecture and the landscape. In a setting less evocative than the Swedish case mentioned above, the use of materials plays a key role in the definition of the atmospheric qualities of the Igualada Cemetery. Miralles and Pinos use concrete and corten steel to emphasize a deliberate confrontation between the roughness and artificiality of these materials and their manipulation as subtle and naturalized components of a new landscape. In both cases, the architectural operation reveals a clear intent to mingle the world of the deceased with that of the living.

In these cemeteries memory and recollection contribute to gain awareness on the present, and project the future. Further, both in Stockholm and in Igualada, the extremely careful detailing of the project suggests a sharp confrontation with the natural, thus asserting the presence of design as a worldly feature. The spiritual evanescence suggested by Nature is thus balanced with craftsmanship to bring forth the telluric character of the everyday. In the projects featured ahead, these negotiations between such aspects as architecture and landscape, the sacred and the worldly are inescapable features that pervade projects designed for such different contexts as Spain, Italy, Belgium, India, and Japan.

The Telluric and the Evanescent

The extension of Gubbio Cemetery in Italia, designed by Andrea Dragoni Architetto, shows a clear example of this attempt to negotiate the telluric with the evanescent. The architect deliberately plays with a composition of solids and voids that frame the per-

斯德哥尔摩的林地公墓,瑞典, Gunnar Asplund 与Sigurd Lewerentz
The Woodland Cemetery in Stockholm, Sweden by Gunnar Asplund and Sigurd Lewerentz

瞬息与永恒

由Andrea Dragoni Architetto设计的意大利古比奥公墓的扩建工程是一个明显的例证,能够证明建筑师想要调和瞬息与永恒的意图。建筑师有意将实体与虚无相结合,构建出一种现实感。一方面,公墓的建筑材料和构成突出了它的庄重和世俗的一面。比如,街道与广场的建造,比如石灰华的使用,表明了它与周围建成景观的直接联系。另一方面,整个工程布局中的天井设计使得唯一的上空空间变成了头顶的天空,巧妙地将现实世界转换为"无形的统治"。然而,这种与世俗的分离与每个庭院中不可或缺的艺术品都构成了一种对比。Dragoni引用William Lethaby的话说,这些建筑空间与建筑材料的特点正好能够让人更好地理解这个世界,而这一点必须要与之分离方能实现。

由Battle i Roig为西班牙的Sant Joan Despí市设计的Sant Joan Despí新殡仪馆就把景观当成了一个重要的建筑因素,这个建筑设计旨在将新建筑与原有的景观相融合。据建筑师本人描述,比如,屋顶的形状就貌似周围景观的延展。这种做法使建筑物与景观间产生很强的依赖性。然而,这种依赖关系并不是直线的。连续的屋顶被断开,展示出由水泥墙、考顿钢柱、窗框及玻璃所围绕的空间的人造属性。穿入屋顶的天井强调了这种由设计师精心布置的建筑空间与容易消散的自然光之间的有趣关系。

重塑边界

景观在殡仪类建筑中的重要性在由Daniel Dethier领导的设计团队所设计的韦尔肯拉特丧葬中心中有更进一步的体现。这个公墓中心的设计旨在在建筑、景观与人之间创造一种有意义的联系。巨大的屋顶重塑了建筑形态,并且与公墓的地下设施相互协调。尽管这些区域的边界是模糊的,这种建筑策略增强了开放感,并使内部与外部、公共与私有空间相互独立。大屋顶将整个建筑延伸,空间特点也是多样的。建筑师们发现,这使得在和谐与沟通之间达成的平衡变得更具挑战性。

ception of reality. On the one hand, the spatial and material composition of the cemetery accentuates its gravity and earthly dimension. For example, the creation of streets and squares, and the use of travertine, suggest direct connections with the surrounding built landscape. On the other hand, the layout of the project, creating courtyards where the only void is framing the sky above, insinuates a shift from the telluric to what the architect calls "the reign of the invisible." This detachment from the worldly is nevertheless contrasted with the inescapable presence of artworks in each courtyard. Paraphrasing Wiliam Lethaby, Dragoni asserts the spatial and material characteristics of these spaces cater for the possibility to understand the world, which is only possible through a detachment from it.

The landscape is also a major factor in the atmospheric qualities of the Sant Joan Despí New Funeral Home designed by Battle i Roig for the Spanish town of Sant Joan Despí. The design of the building was chiefly determined by an attempt to adapt the new construction to the existing topography. The shape of the roofs, for example, appears as an extension of the surrounding landscape, as the designers suggest. This operation produces a strong interdependence between the building and the site topography. This relation, however, is not linear. The continuity of the roof is broken to reveal the artificial nature of the spaces enclosed by concrete walls, corten steel pillars and window frames, and glass. The light wells pierced in the roof emphasize this playful relation between the telluric spaces carefully shaped by the designers and the evanescent presence of natural light.

Reshaping Boundaries

The vital importance of the landscape in the architecture of funerals is further expressed in the Welkenraedt Funeral Center, designed by the team leaded by Daniel Dethier. The funeral center was designed as a place to foster the creation of meaningful relations between architecture, the landscape and the people. A large roof reproduces the site morphology and accommodates beneath it the programmatic components of the funerary facility. This design strategy enhances the feeling of openness and produces a strong interdependence between interior and exterior, private and public. The boundaries that separate these realms are blurred, though. Under the big roof that spans over the whole complex, the character of the spaces is diverse, generating, as the architects recognize, a challenging balance between harmony

伊瓜拉达墓园，巴塞罗那附近，西班牙，Enric Miralles和Carme Pinos
Igualada Cemetery, near Barcelona, Spain by Enric Miralles and Carme Pinos

边界条件这一主题，在曼奇尼公司设计的位于哥印拜陀市的G.K.D.慈善信托火葬场建筑中也有所体现。与景观相关的边界定义与我们之前所讨论的建筑有所不同，景观在那些建筑策略中已经被包含，是它们其中的一部分。恰恰相反，在这个案例中，建筑师旨在人口密集的城市里创造一块和谐之地。这个设计中的亭馆布置创造了一种精心策划的空间序列。不同的建筑材料在透明与不透明的封闭空间之间转换。多孔的墙面设计也被运用到了屋顶上，加强了空间的戏剧性，从而迎合了最重要的丧葬惯例。

然而，哥印拜陀市火葬场渗透式的墙壁和屋顶展示了空间之间连续的对话。在桑塞波尔克罗公墓，建筑的墙壁清楚地展现了空间之间的界限。这个建筑项目是由翟尔玛尼建筑协会工作室为静态建筑大师皮耶罗·德拉·弗朗西斯卡的出生城市所设计的。建筑强调了托斯卡纳地貌的特点，构成新墓地以及早在19世纪就已经存在的公墓的抽象外围护结构。长长的砖墙给这里的地形赋予了新的定义。这堵砖墙看上去为群山奠定了一个抽象的基础，也可以作为桑塞波尔克罗公墓的背景。实际上，这个庭院的背景上打了许多的洞，这些洞内陈放着逝者的骨灰盒。建筑整体严格的几何设计以及材料的选取都体现了这个建筑想要突出永恒这一主题的意图。以上所讲述的几个建筑案例是一些不同的殡仪类建筑。一些有象征性的标志被公墓、火葬场或是殡仪馆所围绕。这些都表明了设计技巧的发展。这些设计技巧旨在在生与死、物质与精神、有限与无限之间建立一座桥梁。在这些完全相反的因素之间进行调和，建筑设计则表现为将自然与人造、返璞与升华相结合。人类在地球上短暂的停留与建筑所营造的一种承载记忆与敬意的氛围不期而遇，它们超越了时间与空间的限制。

and communicative strength.

The definition of porous boundaries is a theme that can also be observed in the Crematorium for G.K.D. Charity Trust in Coimbatore designed by Mancini Enterprises. The relevance of the landscape for the definition of these boundaries is, however, different from most of the cases discussed above, where the natural landscape was part and parcel of the design strategy. In this case, conversely, the project aims at creating an island of harmony in a dense part of the city. The layout of the pavilions in the plot creates an elaborate sequence of spaces, whose limits alternate between opaque and permeable enclosures, emphasized by different material options. The porosity of the walls is also transported to the roof enhancing the dramatization of the spaces that accommodate the most important funerary rituals.

Whereas in the crematorium of Coimbatore the permeability of the walls and roof suggested a continuous dialogue between spaces, in the Sansepolcro Cemetery the walls clearly define the limits between the spaces enclosed by them. The project, designed by Zermani Associati Studio di Architettura for the city that was the birthplace of Piero della Francesca, enhances the characteristics of the Tuscan landscape developing an abstract envelope that encloses both the new burial spaces as well as the additions to the existing cemetery that were made from the nineteenth century on. The long brick walls redefine the topography of the site, creating an abstract base that seemingly supports the hills that configure the backdrop of Sansepolcro. This base is, in effect, perforated by a sequence of courtyards structured by brick walls punctuated by ossuary cells. The strict geometry of the whole ensemble and the careful use of materials suggest a deliberate attempt to enhance the timelessness of this architectural operation. The projects discussed above show different approaches to the architecture of funeral. The symbolic charge of the spaces enclosed in cemeteries, crematoriums or funeral homes, fosters the development of design strategies that attempt to bridge the divide between heaven and earth, material and spiritual, finitude or infinite. Negotiating the boundaries between these pairs of opposites becomes thus a vital aspect to develop an architectonic perception of space that combines the artefact with nature and the telluric with the ethereal. The transience of human earthly existence is countered by an architecture that creates atmospheres to accommodate memory and homage, going beyond the limits imposed by time and space. Nelson Mota

殡仪类建筑 Funeral

古比奥公墓的扩建
Andrea Dragoni Architetto

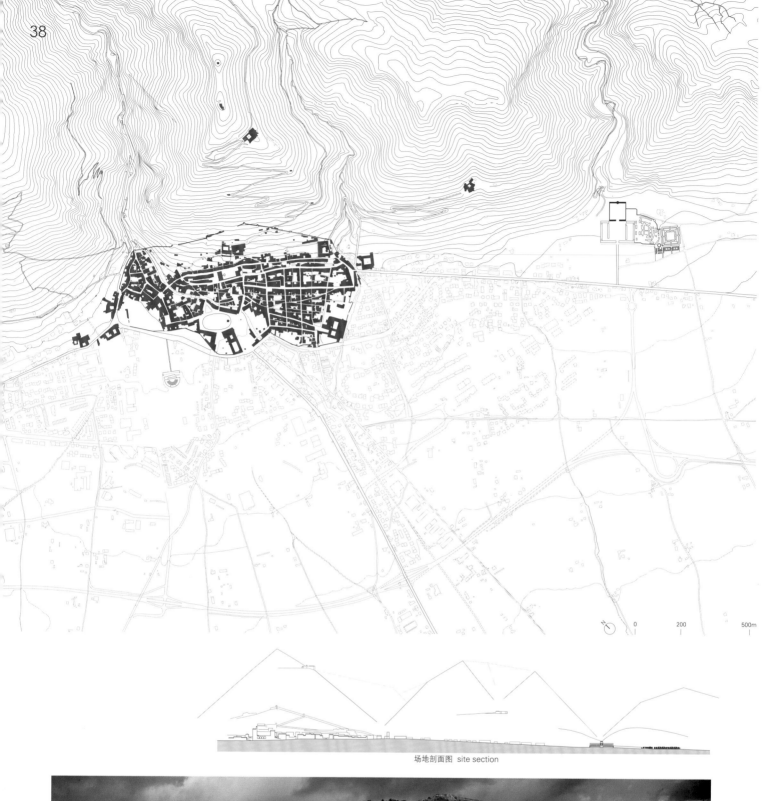

场地剖面图 site section

古比奥公墓的扩建是公共建筑新模型研究的结果。一方面，它使意大利最重要的中世纪城市之一——古比奥市的纪念性建筑进入了新阶段，另一方面，它旨在重新定义自己，以及自己在城市里的中心位置。这个建筑规划由线性体量的街区构成，其排布方式反映了以周围的景观和历史之城为特点的乡村布局。都市设计理念是通过将大广场的外围护结构包含在内来体现的，这样能使建筑结构更加有空间韵律感。

这些空间的设计灵感来自于詹姆斯·特瑞尔的作品《Skyspaces》，旨在设计一处令人愉悦的公共空间。该空间独立于公墓之外，给人提供一个能够停下来思考的地方，这就是立体的"沉默的广场"，有着开放的天花板，窗子始终对着天空。

透过窗子看到的天空打开了人们的思维，他们的视野和思绪可以放任驰骋，不受地球母亲的重力束缚，人们的维想可以更加有空间感，而且上升到精神层面。

建筑与天空的关系实际上是想诠释空间和时间。通过这种方式，人们可以重拾自我。建筑的空间推向天空，喻意着天堂的界线，这是现代城市中我们生命的最后一个层面。与此同时，它朝向天空，重新诠释了利昂纳·巴蒂斯塔·阿尔贝蒂的窗子，这扇窗子被文艺复兴时期的建筑师想像成一个门槛，它是唯一一个能够保持平和的人造建筑设计。它使我们的灵魂回到泰然自若的状态，否则，人们将无法战胜困难。

这些沉默广场的氛围通过一系列永恒特定的人造装饰显得更加暗示性，这些人造装饰捕捉到了从黎明到黄昏光影的变化效果。这些装饰由两名著名的意大利艺术家完成(Sauro Cardinali及Nicola Renzi)。他们的合作始于建筑项目的初级阶段。这些装饰与建筑息息相关，它给沉默空间以及城市里人们的冥想赋予了新的空间。

William Richard Lethaby认为人们并不能完整地认识这个世界。他们必须要先跳出来，人们只有脱离这个世界，才能理解这个世界。

这样的话，一座建筑就可以被看作是一个世界的模型。它代表了我们在这个世界上并不能经历的一种顺序。但同时，作为存在于这个世界的一座建筑，它是可以被感知的。

Gubbio Cemetery Extension

The enlargement of the Gubbio Cemetery is the result of studies of a new model of public building. On the one hand, it has developed the latest phase of growth of the monumental cemetery in Gubbio, one of Italy's most important medieval cities. On the other hand, it intends to redefine its meaning and centrality within the structure of the city. The plan is in an urban structure consisting of linear stereometric blocks arranged in such a way as to reflect the rural layouts that characterize the surrounding landscape and the historic city. This concept of urban settlement is emphasized by the inclusion of large square enclosures designed to be open spaces that provide the structure with spatial rhythm. These spaces were inspired by James Turrell's Skyspaces and are designed to be enjoyable public areas, independently from the cemetery, offering an opportunity to pause and reflect. These are cubic "squares of silence" having open ceilings that evoke windows open to the sky.

The sky thus framed opens the mind to the reign of the invisible, allowing sight and thought to abandon Mother Earth's gravity

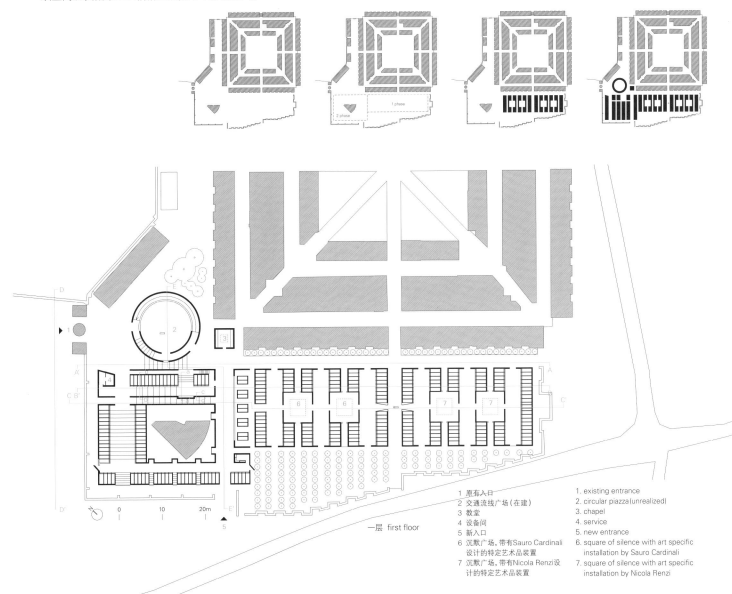

一层 first floor

1. 原有入口 — 1. existing entrance
2. 交通流线广场(在建) — 2. circular piazza (unrealized)
3. 教堂 — 3. chapel
4. 设备间 — 4. service
5. 新入口 — 5. new entrance
6. 沉默广场，带有Sauro Cardinali设计的特定艺术品装置 — 6. square of silence with art specific installation by Sauro Cardinali
7. 沉默广场，带有Nicola Renzi设计的特定艺术品装置 — 7. square of silence with art specific installation by Nicola Renzi

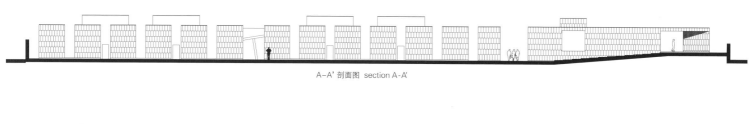

A-A' 剖面图 section A-A'

B-B' 剖面图 section B-B'

and acquire a more aerial and spiritual dimension.

This relationship with the sky intends to define space that is also time, in such a way that you can find yourself again, a space that thrusts the horizon upwards like a metaphor of the boundaries of heaven, the last horizon of our life in a modern city. At the same time, opening to the sky, it re-interprets Leon Battista Alberti's window, a window that is like a threshold, imagined by the great Renaissance architect as the only architectural artifice able to "instil the peacefulness" evoked by the celestial void that, descending from above, takes us back to the imperturbable state of the soul without which overcoming the adversities of life is impossible.

The atmosphere of these "Squares of silence" is made more suggestive by a series of permanent site-specific artistic installations that capture the changing effects of light and shadow from dawn to dusk. These installations were created by two important Italian artists (Sauro Cardinali and Nicola Renzi), with whom collaboration began during the initial stage of the project. This contribution, strongly linked with architecture, helps to define a new space for silence and meditation within the city.

William Richard Lethaby said that human beings cannot understand the world as a whole. They must first move away from it, and only after having achieved this detachment can they achieve understanding.

In this sense a building can be seen as a model of the world; it represents an order we cannot directly experience in the world, but at the same time it makes perceptible, within the limits of a building, which exists in the world.

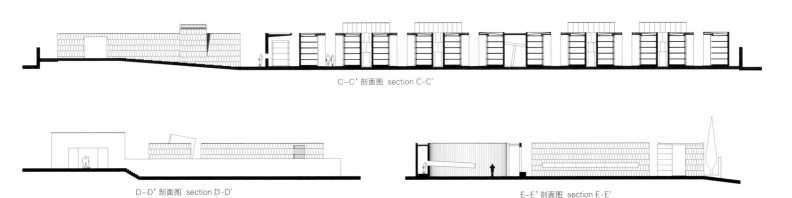

C-C' 剖面图 section C-C'

D-D' 剖面图 section D-D'

E-E' 剖面图 section E-E'

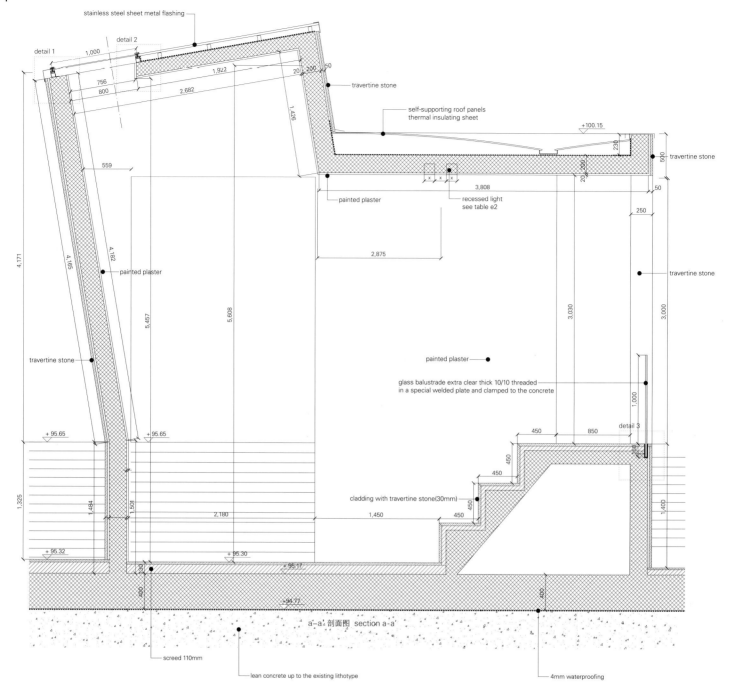

a-a'剖面图 section a-a'

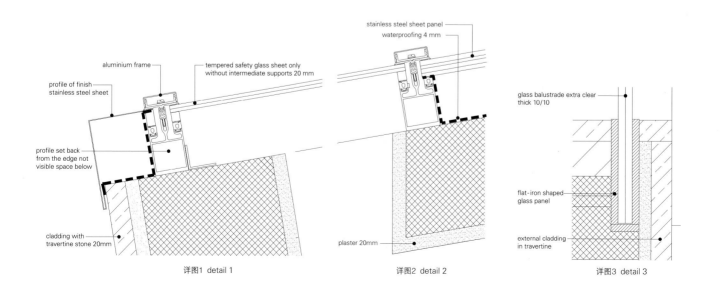

详图1 detail 1

详图2 detail 2

详图3 detail 3

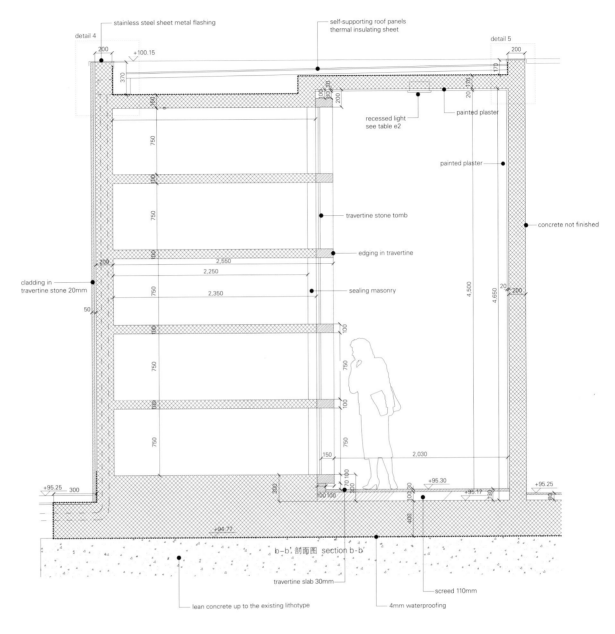
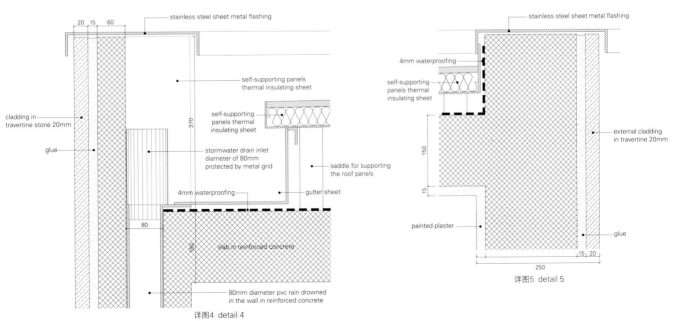

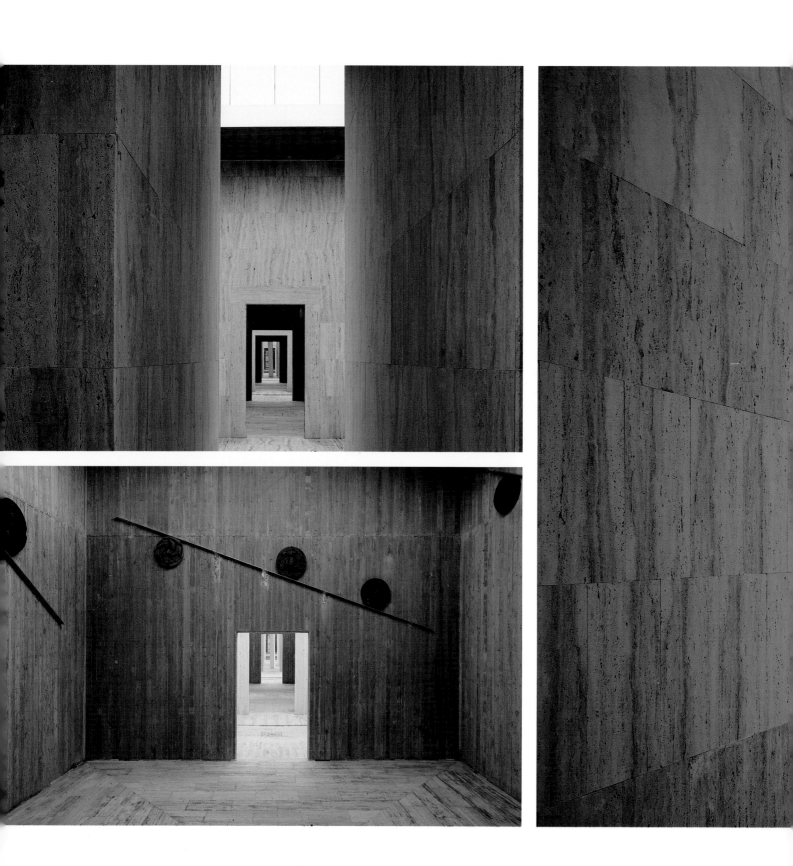

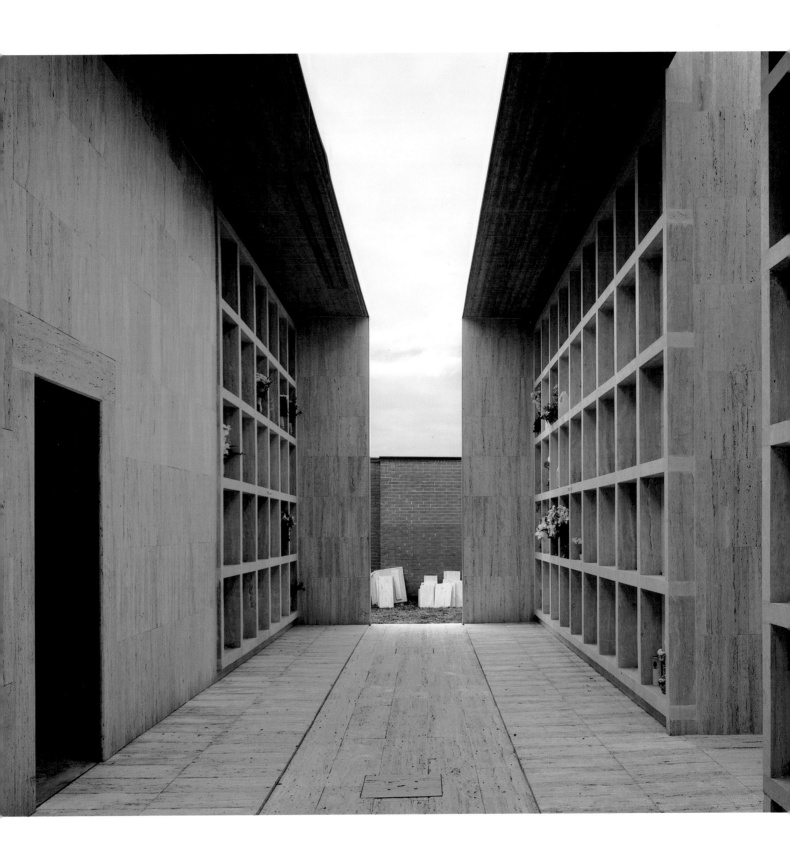

项目名称：Cemetery Complex
地点：Via del Crocifisso, Gubbio, Italy
建筑师：Andrea Dragoni Architetto, Francesco Pes
项目团队：Andrea Moscetti Castellani, Giorgio Bettelli,
Michela Donini, Raul Cambiotti,
Antonio Ragnacci, Cristian Cretaro, Matteo Scoccia
艺术家：Sauro Cardinali, Nicola Renzi
结构设计：Giuseppe Artegiani, Marco Bacchi
植被设计：Italprogetti
安全协调：Claudio Pannacci
项目主管：Francesco Pes, Paolo Bottegoni
模型：Giuseppe Fioroni
甲方：Comune di Gubbio
表面面积：4,035m² 建筑面积：1,800m²
总体积：6,000m³
设计：2004—2005 施工时间：2005—2011
摄影师：
©Alessandra Chemollo_ORCH(courtesy of the architect) - p.36~37, p.38, p.40~41, p.42, p.43, p.46, p.47, p.48~49
©Massimo Marini(courtesy of the architect) - p.50, p.51

殡仪类建筑 Funeral

韦尔肯拉特丧葬中心

Dethier Architecture

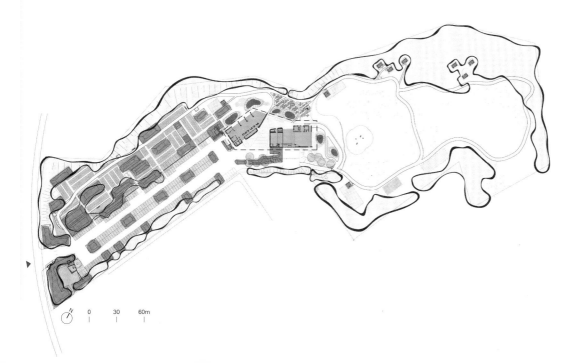

0 30 60m

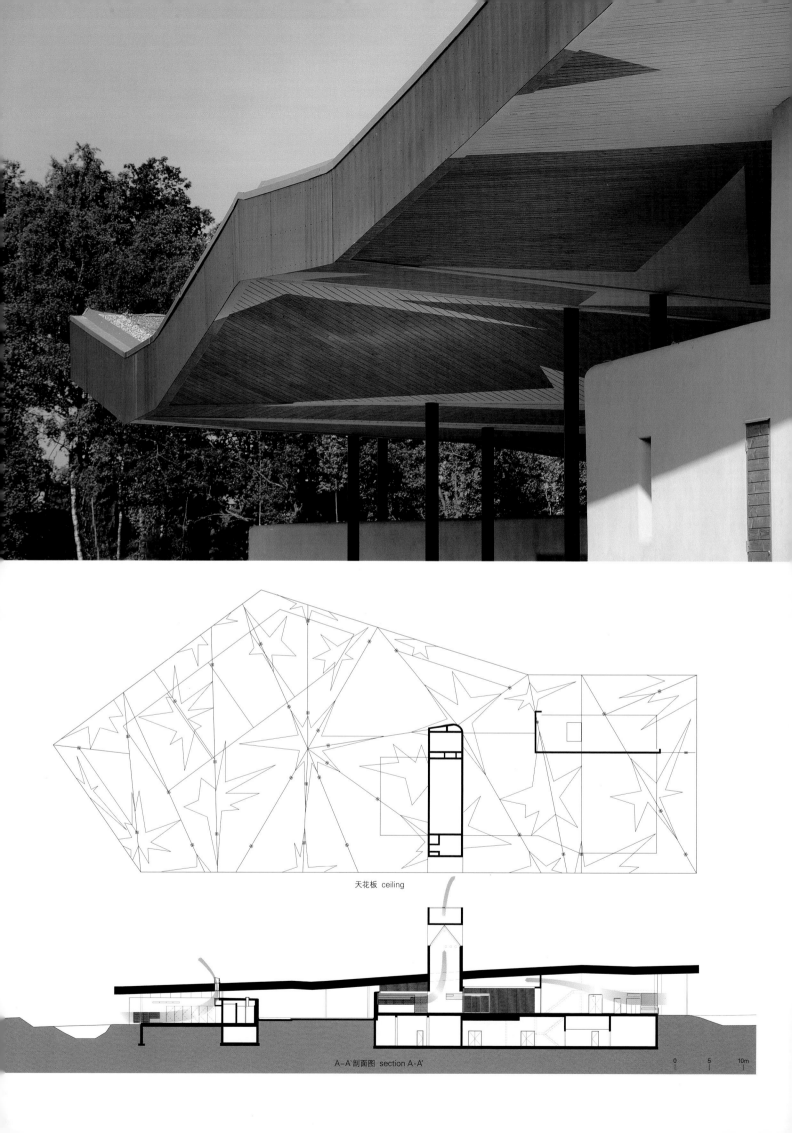

天花板 ceiling

A-A'剖面图 section A-A'

和谐的构成

韦尔肯拉特丧葬中心由两个主要建筑结构组成，它们由种着茂盛植物的屋顶连接。稍小一点的建筑内有一个自助餐厅，其中包含就餐区。大一些的建筑的一层建有一座小礼拜堂，楼上是办公室及技术区，地下是火化设备。建筑的入口设在一侧，保安室将建筑的边界线及主楼融合在一起。此建筑的设计非常注重容积及相互关联的和谐性。水泥墙的砌成赋予它一种平衡感和安全感。门框及墙基上的曲线设计是这座建筑所偏好的一种表达方式。这种方式所带来的一种心神稳定的感觉，与主楼上的塔所表现的沟通的力量并不抵触。这座塔主要是用来遮盖火化设备的烟囱，同时还给小礼拜堂提供自然通风以及使建筑的中心位置得到光照。

自然的延伸

这个建筑的基础之一便是将建筑与自然环境相融合。因此，这座建筑包含了很多大落地窗的设计。这个转换理念是建筑要素的关键。从外观来看，这座建筑像是从地里长出来的。从建筑的屋顶也可以看出，屋顶的"舒展"与地势是和谐的，延续了建筑的足迹，给这座建筑提供更多的室外区域，与周围的环境相融合。因此，景观依靠建筑得到延伸，它整合了周围的外部空间。丧葬中心所在的自然区占地超过六公顷。丧葬中心占地三公顷。这一区域是由景观建筑师Erik Dhont精心设计开发的。一进入中心，交通流线按照功能性来安排，同时考虑到人对它的感知。两边种植着多种当地植物的道路将访客引入停车区。此外，整个景观的设计构思是想涵盖更多的功能（如：公墓、分散的草坪、休息区、火葬场以及保留最初自然景致的生活区）。每一个区域以一种简单、自然、"平静"的方式存在。景观中还精心设计了各种各样的墓地、相应设备及标识。这些都与Nicolas Kozakis艺术集成的设计方法相吻合。这种设计方法是他在建筑职业生涯一开始就主张的，那个时候Daniel Dethier便邀请他加入设计过程中。他的加入使自己设计的特点在丧葬中心的天花板及焚尸炉壁上的设计都有体现。

Welkenraedt Funeral Center

A Harmonised Composition

The Welkenraedt Funeral Center consists of two main structures that interconnect beneath a broad planted roof. The smaller of the two contains a cafeteria and associated dining areas, while the larger structure houses chapels on the ground floor, with offices located on the upper floor, and technical areas and cremation facilities in the basement. At the entrance to the site, set off to one side, a caretaker's residence blends with both the contours of the site and the formal aspects of the main building. Particular attention has been paid to the harmony of the volumes and their interconnections. The result – a feeling of balance and reassurance – is reinforced by the finishing of the concrete walls. Curves are a preferred means of expression, including in doorframes and the bases of walls. The tranquillity this affords is not antithetical to the communicative strength of the tower that emerges from the central building. The tower is used to conceal the chimneys of

项目名称：Welkenraedt Funeral Center
地点：Welkenraedt, Belgium
建筑师：Daniel Dethier
艺术家：Nicolas Kozakis
景观建筑师：Erik Dhont
室内面积：3,290m² 室外面积：28,944m²
竣工时间：2012
摄影师：©Serge Brison & Thomas Faes (courtesy of the architect)

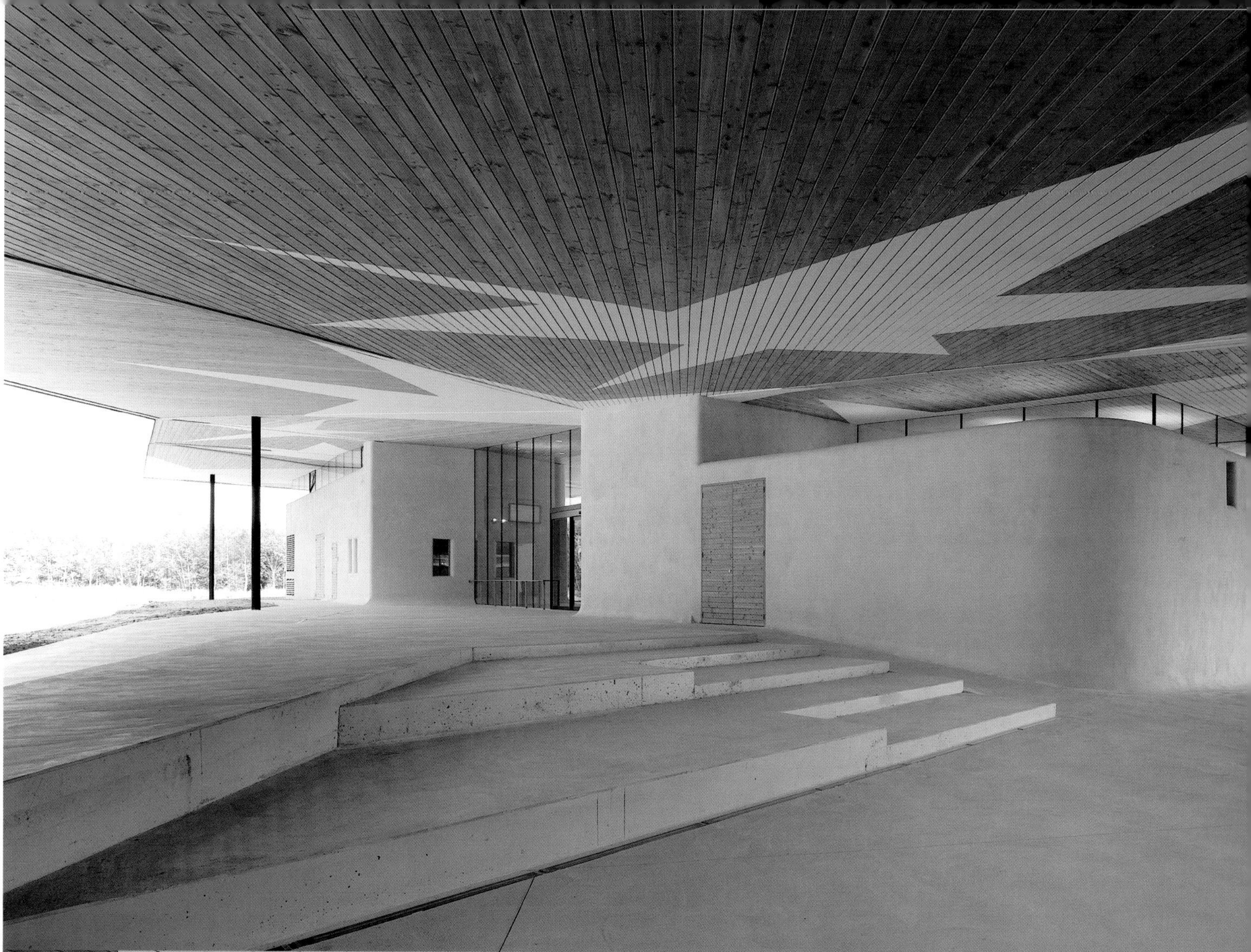

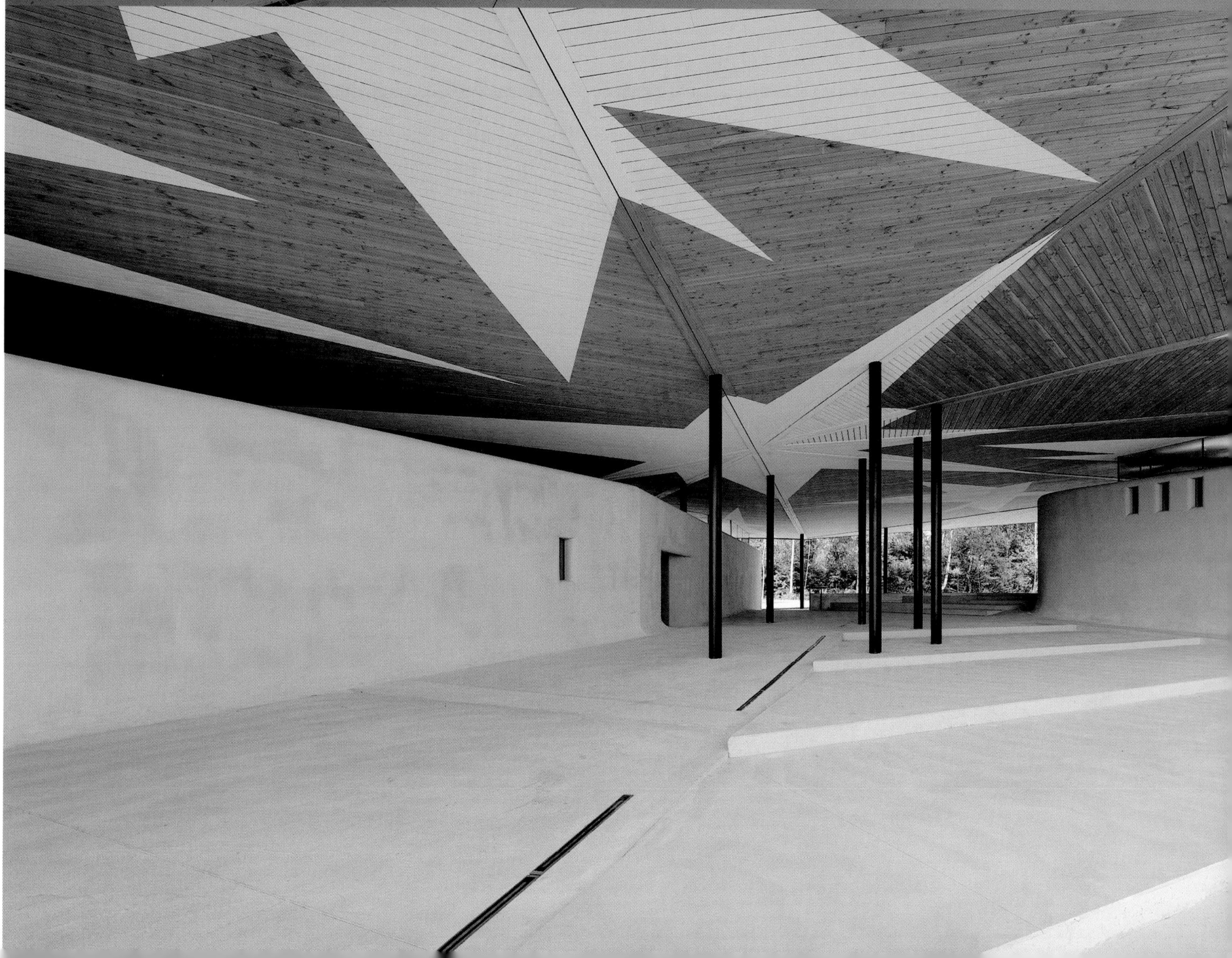

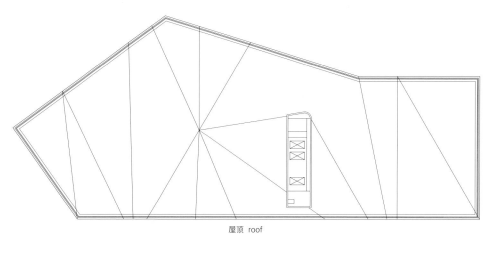

屋顶 roof

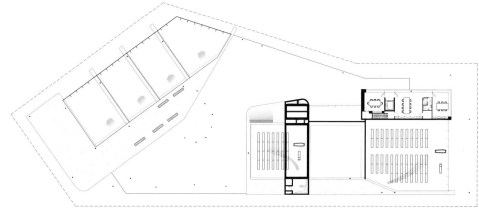

二层_通风 second floor_ventilation

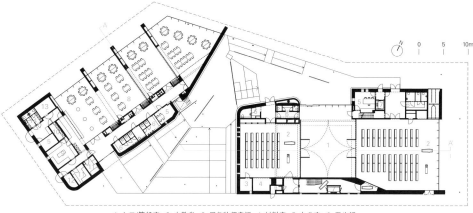

1 入口/等候室 2 小教堂 3 司仪牧师房间 4 材料库 5 办公室 6 卫生间
7 电梯 8 自助餐厅 9 厨房 10 冷藏室 11 洗衣店 12 技术室 13 仓库 14 衣帽间
1. entrance/waiting room 2. chapels 3. officiant's room 4. material storage 5. office 6. WC
7. lift 8. cafeteria 9. kitchen 10. cold room 11. laundry 12. technical room 13. stock 14. locker rooms
一层 first floor

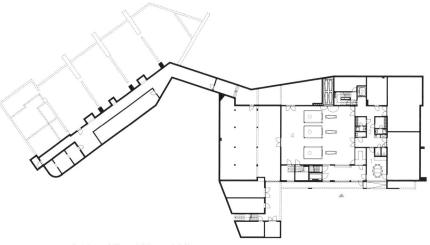

1 办公室 2 电梯 3 衣帽间 4 火葬炉 1. office 2. lift 3. locker rooms 4. cremation ovens
地下一层 first floor below ground

the cremation facilities, while providing natural ventilation to the chapels and bringing light down into the building's core.

An Extension of Nature

One of the cornerstones of the project is the integration of the site into the natural environment. This includes the design of the structure, which incorporates vast floor-to-ceiling windows that give onto the outside. The concept of transition is a key architectural element, which is underscored by the impression that the buildings are rising from the earth. It can also be seen in the design of the roof, which "unfurls" in harmony with the topography and continues out over the footprint of the buildings, providing shelter for open-air zones in continuity with the environment. Views of the landscape thus extend throughout the buildings, which organise, reveal and bring together the various peripheral spaces. The natural site where the funeral center is located consists of more than six hectares of land, three of which were built. The site was developed very carefully, working in close collaboration with landscape architect Erik Dhont. Upon entering the center, circulation is organized in a functional, sensitive manner – access roads lined with banks of indigenous species lead visitors to parking areas. In addition, the entire landscape design was conceived to bring together the site's various functions (cemetery, scattering lawns, rest areas, crematorium and the original natural site with its specific biotope) into a carefully-composed whole where each area appears to have been placed there in a simple, natural and "serene" manner. The various types of burial places have also been thoughtfully designed, along with the furnishings and signage.

In keeping with his approach of artistic integration, which he has advocated since the very beginning of his career, Daniel Dethier asked Nicolas Kozakis to join his team in the design process. His participation resulted in design elements that appear on the center's ceiling and on the walls of the cremation ovens.

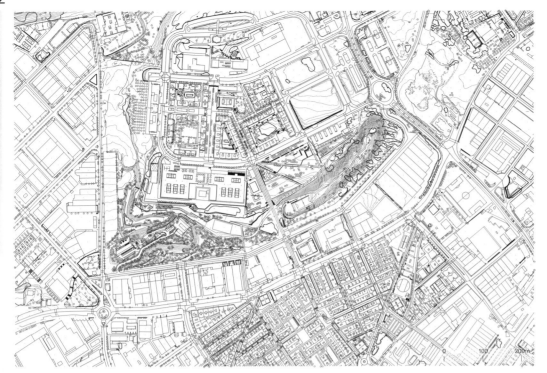

Sant Joan Despí 新殡仪馆

Batlle i Roig Arquitectes

Sant Joan Despí新殡仪馆的设计主张源于对都市介入的理解。它建在康斯坦察公园和已经存在的公墓构成的城市集合体的端部。它所选取的位置符合如下标准：通过殡仪馆的建造改善现存的公墓设备。所采用的建造系统和技术方案将对现有的设备起到补充作用，能推动可持续性发展，并能提高能源使用效率。

建筑师将现有的破败的公墓入口建得体面些。将里面的设施重新整理、布置，设立分化的行人区与停车区。在景观设计理念中保留公园的特质。这个区域尽管位于都市，但仍要保留和重塑新建区域周围的绿色斜坡。将部分建筑与周围原有的地貌进行融合。这个设计主张以现存的公墓为轴。新楼建在北边，南边的停车场与已存道路相连。通往现存公墓的入口轴线成为连接三个重要部分（公墓、殡仪馆、停车场）的广场。这些区域周围原有的植被覆盖面积增加了，以保留场地原有的的公园特点。

场地中的建筑群并不是仅仅与周围原有的地形相融合。该建筑的屋顶便是就地势而建的倾斜屋顶。从周围较高位置的街道上看，部分屋顶上铺设了绿色植被，与周围的绿色斜坡融为一体，改善了建筑的整体外观。此外，这种技巧减少了显露在外的建筑容积，减少了建筑视觉上的大小，增加了绿色面积。

这个建筑的楼层计划占地面积为700m²。建筑的两个区域有非常清楚的区分。一个是公共区域，由一系列房间组成，里面有一些供人们使用的设备；一个是私人区域，主要为顾客提供服务，以减少他们为葬礼进行的准备工作并减少逝者的棺木在两个区域间移动的次数。整个楼层的设计还包括一系列天井。这些天井对建筑空间进行了组织、分层，并提供了照明，而且，给不同氛围的环境提供了一个过渡。

这个建筑的结构体系由墙、用松木制模的钢筋混凝土板以及由扁钢条制成的考顿钢柱建成。因为材料的使用非常简单，所有的这些建筑元素定义了这座建筑的整体形象和特点。天然的石材铺面和使室内具有温暖感觉的垂直木条饰面完善了建筑的物质性。钢柱使光产生了折射，形成了一个视觉上的滤光器，防止建筑内部直接受到光的照射。

裸露的建筑构件纹理相互交融所产生的物质性和自然光一起，限定和确定了每个空间的氛围，伴随着四处访客的哀悼声。这样一来，每一个建筑空间所照到的光都不同于别处，是独一无二的。

Sant Joan Despí New Funeral Home

The proposal for the Sant Joan Despí New Funeral Home arises from understanding the intervention as the end of the urban ensemble formed by the Fontsanta Park and the existing cemetery. Its location is based on the following criteria: Improve the existing cemetery facilities with a new funeral home that will complement the current offered services, using construction systems and technical solutions that will promote sustainability and energy efficiency.

The architects dignify the entrance of the existing cemetery which is quite degraded, organizing, arranging and equipping it with differentiated pedestrian and parking areas. Integrate landscaping in the proposal respects the park character. The place should, given its urban location, preserve and reforest the existing green slopes around the intervention area and incorporate part of the building volume to the existing topography.

The axis of the existing cemetery organizes the proposal, in the north the new building is implanted and in the south a parking lot is connected with the existing roads. The entrance axis to the existing cemetery becomes an access plaza relating the 3 main parts

项目名称：New Funeral Home in Sant Joan Despí
地点：Sant Joan Despí, Barcelona
建筑师：Enric Batlle i Durany, Joan Roig Duran, Albert Gil Margalef
合作者：architect_Miriam Aranda
农业工程师与景观设计师_Dolors Feu
技术建筑师_Diana Calicó, Elisabeth Torregrosa
设备工程师_SJ12, Albert Colomer
结构工程师_Static, Gerardo Rodriguez
面积：700m²
施工时间：2009-2011
摄影师：©Jordi Surroca

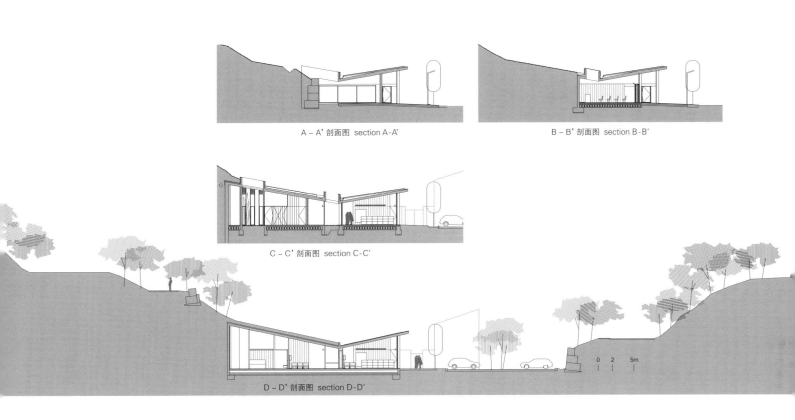

A-A' 剖面图 section A-A'　　　　　　　　　　B-B' 剖面图 section B-B'

C-C' 剖面图 section C-C'

D-D' 剖面图 section D-D'

(cemetery, funeral house and parking lot). Around these areas the existing vegetation is increased in order to complement the park character the site has.

The building integration on site parts from the adaptation to the existing topography, with a set of pitched roofs on the terrain. The vegetation treatment of part of these roofs pretends to fade with the adjacent green slopes and improve the vision of the ensemble from the perimeter streets, on a higher level. With this strategy, in addition, the apparent building volume is reduced, lowering the vision of the construction and increasing the green surfaces.

The 700sq meters floor plan of the building, lays out an organization in two areas clearly differentiated, by a public area composed by a set of rooms designed to serve the users of the facility and a private area composed by the needed service rooms for the deceased preparation and the coffins movement between them. A system of patios completes the layout of the floor plan, and these patios organize, rank and illuminate the spaces and establish filters between different ambiances.

The structural system is composed of walls and reinforced concrete slabs formed with pinewood boards and corten steel pillars made of flat bars. All these elements define the building image and character providing simplicity to the materiality of the piece. The materialization is completed with natural stone pavements and wooden vertical facing producing interior warmth. The steel pillars generate a light gradient, establishing visual filters and protecting the interior from the direct sunlight.

The materiality generated by the assortment of exposed structural element textures together with the natural light qualifies and determinates the atmospheres of each space, accompanying the visitor's mourning at every turn. In this way each space is illuminated by a specific light different from the rest.

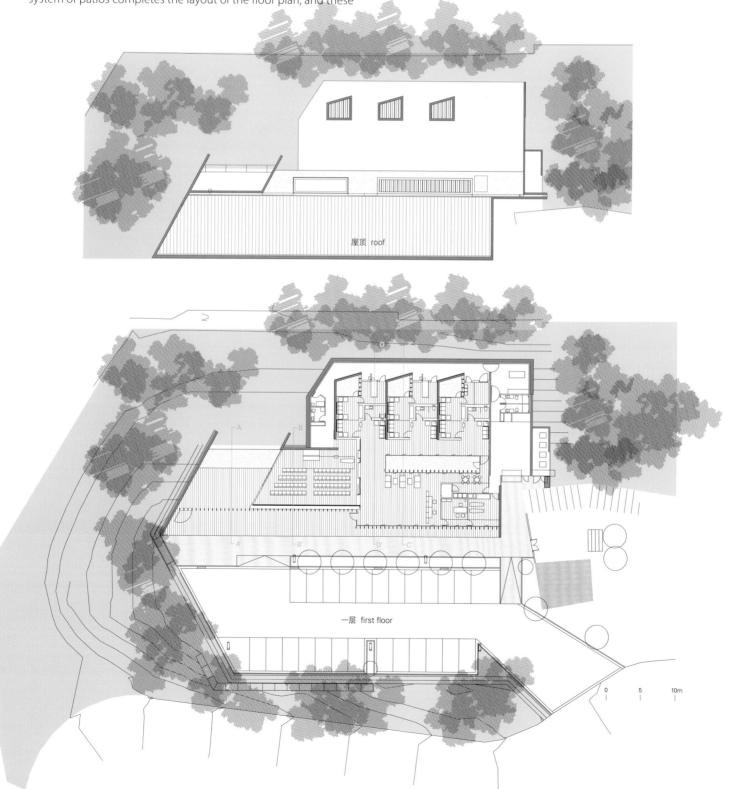

屋顶 roof

一层 first floor

两座用混凝土做饰面的展亭专门在逝者的尸体被送到火化炉之前,为葬礼的最后两个仪式提供场地——这两个仪式往往要有很多吊唁者出席。

第二天早上逝者的亲属从管理楼中将骨灰盒取出,然后把它送到公园里,那里有一些小的展示馆用来举行亲属们特别要求的仪式。

火葬场就设计在了城市人员密集的地方,旨在为人们提供一个简约但不失庄严的环境。它使人们想起在城市的发展阻止这样的火葬仪式之前的传统的火葬举行地——河边。

从传统上讲,火葬场建筑通常是建在城市边缘的河岸上或开阔地带上的展厅结构。人们应该理解这种场景对于参加葬礼的人们的意义。举办葬礼时,周围是大片空地和绿色植物的河边点燃着火葬用的柴堆。然而,在今天的大多数城市,由于城市的扩大,这些用于火葬的柴堆都被城市包围……这也是这个建筑设计的动机所在。花园将两个大厅包围起来,通过选择草、河边的风景这些元素使人们想到河边的火葬的场景。

逝者被放在一个特制的木棺材里,由最亲近的男性亲属抬着。伴着圣歌,这些抬棺者将逝者从家里抬到火葬地。这些逝者现在是由汽车载到火葬场的。从火葬场的入口修建一条道路通往火葬炉的炉腔是非常必要的。对建筑师而言,一进入入口就修建一个临时停靠站,在那里卸下棺木,在火化之前,将它抬到一些仪式指定的地方是一项比较合理的建议。

接着,根据仪式的需要,逝者的身体将被清洗,是要洗去所有存留的欲望,据说这些欲望在死后仍然存在。整个仪式过程在通往火葬炉的路上以及在火葬炉处会暂停一下。在那里,逝者将会被放置在一个平台上,接着圣水会象征性地洒在他/她的身体上,以示清洗。

纵观仪式的整个过程,火葬用的柴堆处于中间位置,喻示着死亡,也表明了人的生命中的任何阶段都不是结束。火葬柴堆要准备些木头,在此期间,会举行一些仪式。接着,火葬柴堆将由死者的丈夫或是长子点燃。

Crematorium for G.K.D. Charity Trust in Coimbatore

2 large pavilions in form finished with concrete provide space for the last rites to be performed – often in attendance of large amount of mourners – before the body is taken to the furnace area. The next morning family members collect the ashes from the admin building and proceed into the garden where smaller pavilions shelter the required rituals.

The design of the crematorium located in a dense part of the city is concerned with providing a humble but dignified environment evoking traditional landscape setting of the river banks where cremations used to be performed before the growth of the city made these impossible.

哥印拜陀市的G.K.D.慈善信托火葬场
Mancini Enterprises

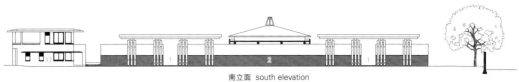

南立面 south elevation

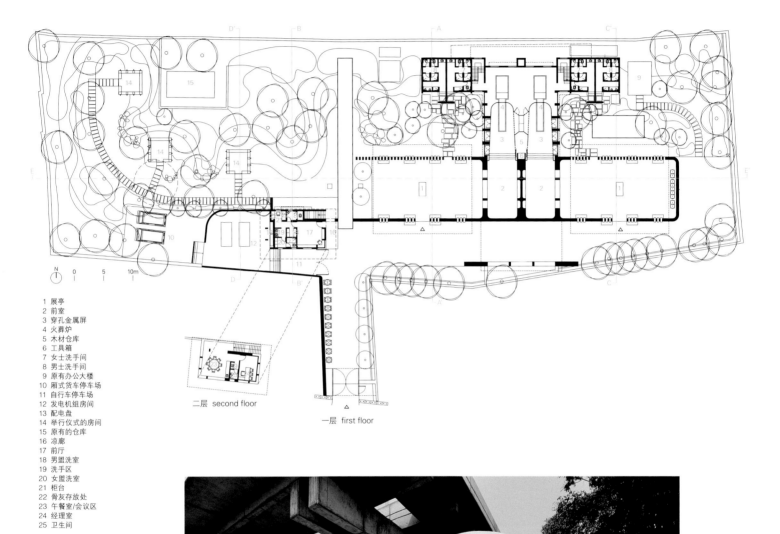

二层 second floor

一层 first floor

1 展亭
2 前室
3 穿孔金属屏
4 火葬炉
5 木材仓库
6 工具箱
7 女士洗手间
8 男士洗手间
9 原有办公大楼
10 厢式货车停车场
11 自行车停车场
12 发电机组房间
13 配电盘
14 举行仪式的房间
15 原有的仓库
16 凉廊
17 前厅
18 男盥洗室
19 洗手区
20 女盥洗室
21 柜台
22 骨灰存放处
23 午餐室/会议区
24 经理室
25 卫生间

1. pavilion
2. ante space
3. perforated metal screen
4. furnace
5. firewood storage
6. tool kit
7. ladies' toilet
8. gents' toilet
9. existing office building
10. van parking lot
11. bike parking lot
12. generator set room
13. electric panels
14. ritual pavilion
15. existing store
16. veranda
17. front office
18. gents' room
19. wash area
20. ladies' room
21. counter
22. ash lockers
23. lunch/conference area
24. manager's cabin
25. toilet

Traditionally, the cremation place is marked by a pavilion structure on river banks or open grounds on the fringes of the city. One should understand the kind of metaphysical importance the scene has in the minds of the people attending the cremation when a funeral pyre is lit in the banks of a water body surrounded by vastness of space and greenery. However, in most cities today, these pyres are encapsulated by growing cities… this is motivation for the design to integrate a garden around the two halls evoking the presence of this important landscape by choice of grasses, driver riverbed elements.

The dead are carried on a specially constructed timber frame by the closest male relatives. These pall bearers carry the dead from their house to cremation grounds to the accompaniment of chants. The dead are now brought to the cremation grounds by motor vehicles. It is essential for a cremation route to be created from the entrance to the furnace chambers. In the architects' case, it would be advisable to create a drop off point just after the entrance from where the body could be carried through a series of spaces for rituals before it actually enters the furnace.

The body is then ritually cleansed to symbolically pacify the remaining desires of the personality, which are said to remain even after the body dies. The procession pauses at the juncture of cremation route and the cremation chambers, where the body is placed on a platform and holy water is symbolically sprinkled on it to signify the cleansing.

In the linear succession of events the funeral pyres occupy an intermediate position. Signifying the idea that death or any stage in life is not an end in itself. The funeral pyre is prepared in wood and more rituals are performed. The pyre is then lit by the husband or eldest son.

项目名称：Crematorium for G.K.D. Charity Trust in Coimbatore
地点：Coimbatore, India
建筑师：Mancini Enterprises Pvt. Ltd.
项目团队：Niels Schoenfelder, J.T. Arima, Bharath Ram K., Ganesh V., Priyanka Rao, Priyanka Bobal, Sridharan A., Rijesh K., Divya K.N.
结构工程师：Mithran structures pvt. ltd.-Coimbatore
景观建筑师：Mancini Enterprises Pvt. Ltd.
土木工程承包商：Ramya Associates-Coimbatore
甲方：G.K.D. Charity Trust
用地面积：4,856m²
建筑面积：1,120m²
施工时间：2012.3~2014.1
竣工时间：2014.1
摄影师：Courtesy of the architect

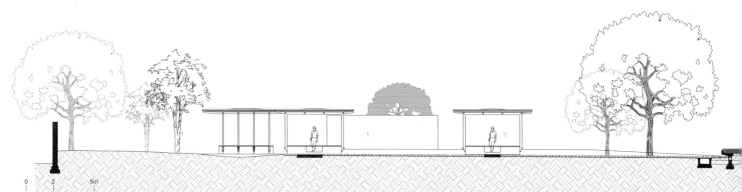

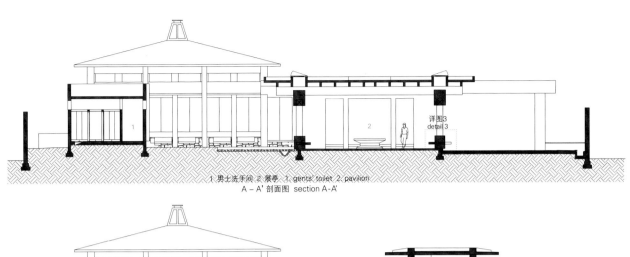

1 男士洗手间 2 展亭 1. gents' toilet 2. pavilion
A－A' 剖面图 section A-A'

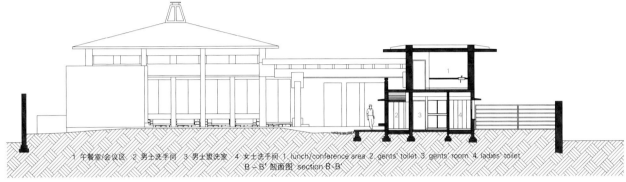

1 午餐室/会议区 2 男士洗手间 3 男士盥洗室 4 女士洗手间 1. lunch/conference area 2. gents' toilet 3. gents' room 4. ladies' toilet
B－B' 剖面图 section B-B'

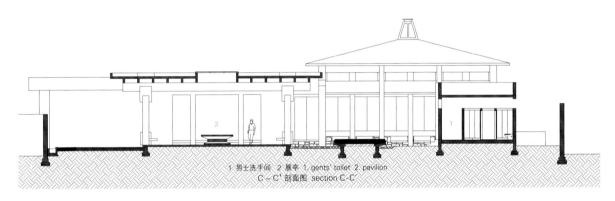

1 男士洗手间 2 展亭 1. gents' toilet 2. pavilion
C－C' 剖面图 section C-C'

1 发电机组房间 2 举行仪式的房间 1. generator room 2. ritual pavilion
D－D' 剖面图 section D-D'

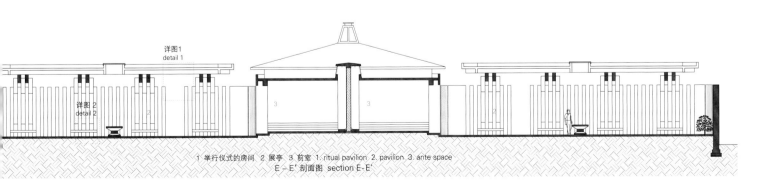

1 举行仪式的房间 2 展亭 3 前室 1. ritual pavilion 2. pavilion 3. ante space
E－E' 剖面图 section E-E'

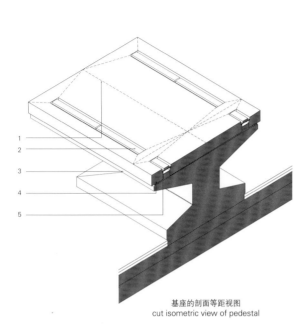

基座的剖面等距视图
cut isometric view of pedestal

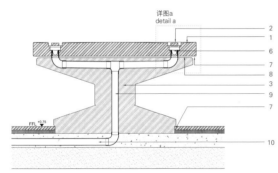

详图2 detail 2

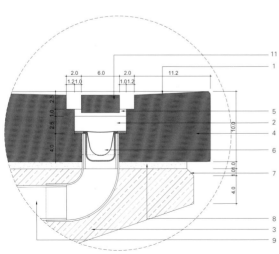

详图a detail a

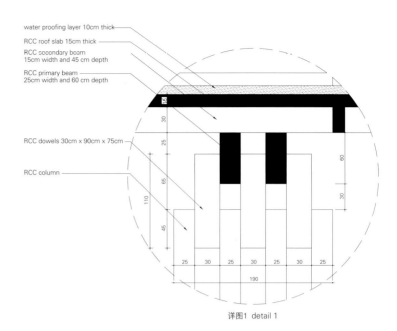

详图1 detail 1

1. stone platform sloped towards drain
2. drain for water run off - 10cm wide
3. RCC platform - 115cm x 225cm x 70cm (exposed concrete finish)
4. polished granite stone platform - 120cm x 230cm
5. steel flat 1cm x 0.6cm
6. CCT-SFC-76 floor trap with gratings - 76cm x 76cm
7. RCC pedestal edge chamfered 1cm x 1cm
8. cement mortar (1:5 ratio)
9. 5.1cm ø drain pipe for water runoff
10. drain pipe take out through PCC floor bed
11. granite stone slat 6.0 x 2.5 cm above drain

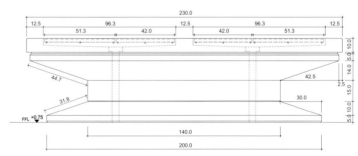

基座立面图 pedestal elevation

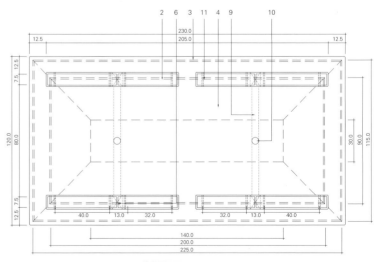

基座平面图 pedestal plan

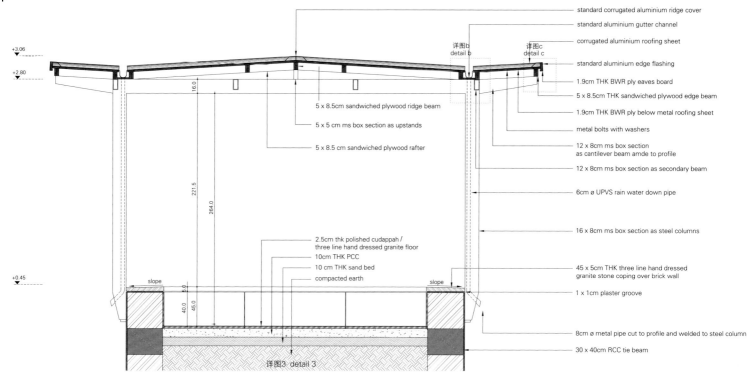

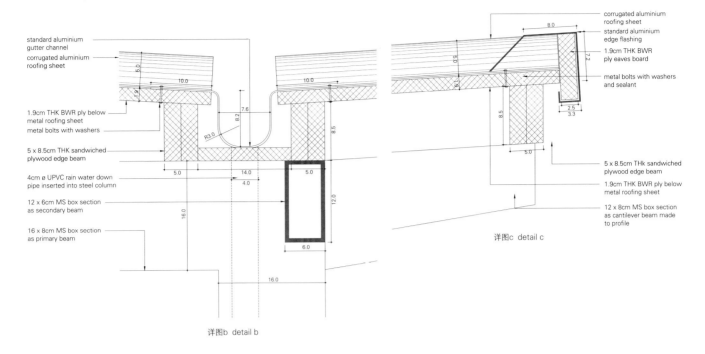

详图b detail b

详图c detail c

殡仪类建筑 Funeral

桑塞波尔克罗公墓
Zermani Associati Studio di Architettura

桑塞波尔克罗位于圣地亚哥及耶路撒冷两地之间的中心位置。这里群山环绕，位于托斯卡纳及翁布里亚之间，也是画家皮耶罗·德拉·弗朗西斯卡的出生地，故此成为了画家作画的地方。皮耶罗经常在室内观察景观，在他看来，背景非常重要。他利用这样独到的观点来把握眼睛、建筑或是纪念碑，以及景观间清晰的关系。对于桑塞波尔罗景观的描述最早是蒲林尼提出的："美丽是景观的一个方面，你可以把它想象成是一座只有自然能够创造出的巨大的圆形剧场。"

新的公墓的外形是一个矩形，它基本上包含了原有公墓的南侧，也有部分是北侧。与19世纪的公墓相比，新公墓添加了很多东西。

建筑的周边是砖砌的阶梯，这是为了适应由东向西高差约为10m的地势变化。阶梯在最高处竖直向上，成为单层立面。从外面看，长达150m的公墓看上去像是群山的基础。从外界看，人们看到的是建筑作为基础支撑着景观；从建筑内部看，人们可以看见周围环绕的自然景观。

建筑内部有一个很大的十字形交叉悬吊结构用来存放骨灰。该结构的顶端面向周围的景观和城市，引入了一条城市通道，指明了新入口的位置。为了凸显周围的墙体，这个十字交叉结构除了较高的安放位置和其所包含的骨灰盒外，什么也没有。

项目名称：Cimitero di Sansepolcro
地点：Cemetery of San Sepolcro, Via Osimo, San Sepolcro(Arezzo)
建筑师：Zermani Associati Studio di Architettura
项目团队：Paolo Zermani, Siro Veri, Mauro Alpini
占地面积：39,883m²
建筑面积：6,805m²
总楼面面积：9,708m²
施工时间：2000—2011(第一阶段和第二阶段)
项目完全竣工：2025
摄影师：©Mauro Davoli(courtesy of the architect)

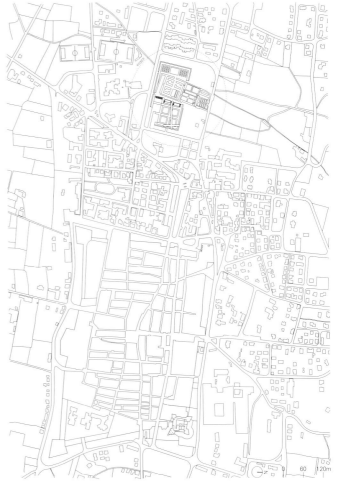

Sansepolcro Cemetery

Sansepolcro, the halfway point between Santiago de Compostela and Jerusalem, and birthplace of Piero della Francesca, is surrounded by the hills on the border between Tuscany and Umbria that the painter transferred into his own pictorial space. Piero often observed the landscape from inside: for him, the background was important, as was the point of view. The extraordinary perspective application of his images imposes the relationship between eye, architecture or monument, and landscape with didactic clarity. The landscape of Sansepolcro is a place recognised as far back as the description given by Pliny: "Beautiful is the aspect of the region: you can imagine it as an immense amphitheatre, such that only nature is able to create."

The new cemetery develops on a rectangular outline, incorporating the existing cemetery completely on the southern side, and partially on the northern side, through various additions, from the 1800s to the present.

The perimetral body, consisting of brick terracing, adapts to the altimetric trends that vary, from the east to the west side, by approximately ten linear metres, but takes the level of the summit of the wall back to a single elevation. From the outside, developing on the longest side for 150 linear meters, the cemetery thus appears to be a sort of base for the hills. From outside, one sees the base supporting the landscape, and from inside one can see the landscape and the cycle.

The large cross, a suspended walkway inside which the ashes are deposited, at the top faces onto the perimeter base, towards the city, introducing a precise urban pathway that indicates the new main entrance. With respect to the surrounding wall, the cross, empty except for the ossuary cells and situated higher up, almost dematerialises.

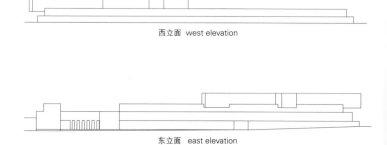

西立面 west elevation

东立面 east elevation

南立面 south elevation

北立面 north elevation

0　15　30m

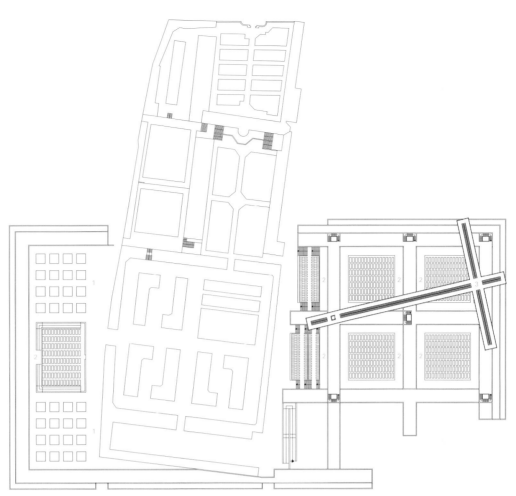

1 教堂 2 墓地 3 藏骨堂 1. chapels 2. burial ground 3. ossuary
二层 second floor

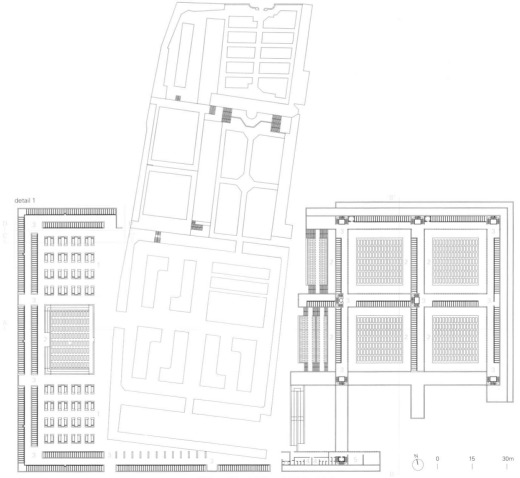

1 教堂 2 墓地 3 坟墓 4 公共浴室 5 仓库
1. chapels 2. burial ground 3. burial in tomb 4. public bathrooms 5. warehouse
一层 first floor

1 坟墓 2 墓地 3 藏骨堂 1. burial in tomb 2. burial ground 3. ossuary
A – A' 剖面图 section A-A'

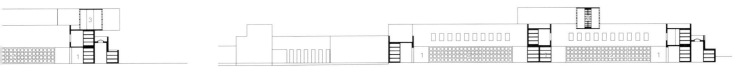

1 坟墓 2 墓地 3 藏骨堂 1. burial in tomb 2. burial ground 3. ossuary
B－B' 剖面图 section B-B'

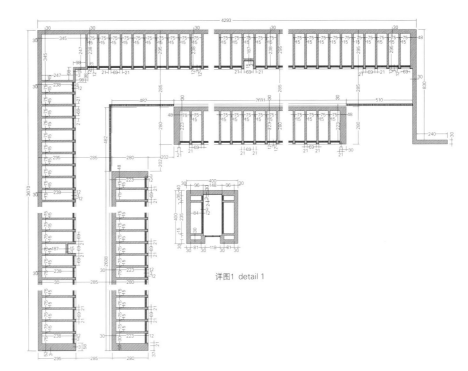

详图1 detail 1

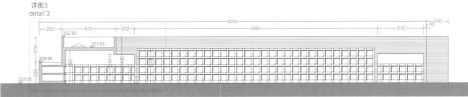

C－C' 剖面图 section C-C'

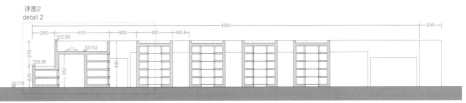

D－D' 剖面图 section D-D'

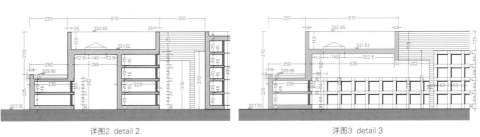

详图2 detail 2　　　　　　　　　　详图3 detail 3

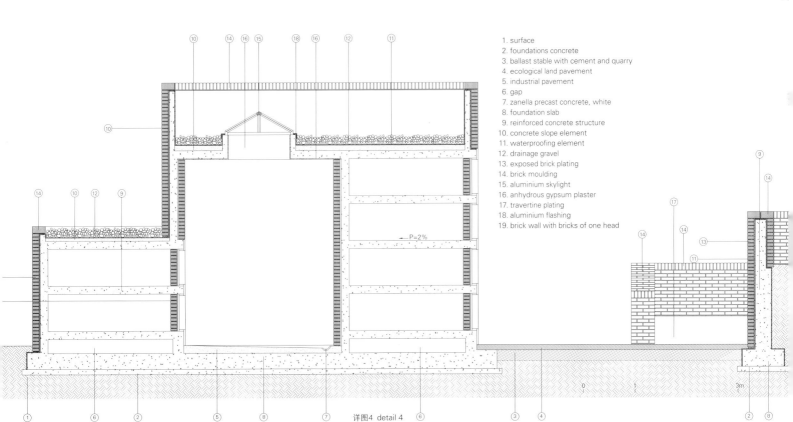

1. surface
2. foundations concrete
3. ballast stable with cement and quarry
4. ecological land pavement
5. industrial pavement
6. gap
7. zanella precast concrete, white
8. foundation slab
9. reinforced concrete structure
10. concrete slope element
11. waterproofing element
12. drainage gravel
13. exposed brick plating
14. brick moulding
15. aluminium skylight
16. anhydrous gypsum plaster
17. travertine plating
18. aluminium flashing
19. brick wall with bricks of one head

详图4 detail 4

社区通道
Gateway to the Com

社区建筑作为桥梁使当地居民和他们的日常生活融为一体，构架起社区条条通道，提供创意和学习空间，成为展示文化和举办社会活动的场所。社区建筑功能多样，建筑设计既能发挥功能性作用也起到了美观作用，与周边环境相得益彰，毫无疑问成为标志性建筑，起到联结社会和社区的纽带作用。社区建筑能在某种程度上改变人与人之间的交往和人们对公共空间的使用，进一步改善社区，为当地居民提供帮助，公布新闻和宣布大事，传承传统和延续历史。社区通道是社区共享之地，宛若灯塔。

那么，社区建筑在空间的设计上有多少发言权呢？新建的共享空间是如何影响社区的整体性的呢？

新规划的空间、崛起的建筑和综合一体化的城市规划、活动和大事件如何影响社区居民个体的生活，让他们有亲切舒适感，同时获得感悟呢？我们如何在社区实现可持续、受欢迎的通道设计呢？

借助科技创新，依托原创观点，融合可持续性特点和理性的风险考虑，建筑可以展现出吻合建筑场地的社区特征。社区建筑代表社区，自然要具备服务当地居民和社区的性能。社区居民必须要对设计产生认同感，同时任何社区项目的实施必须通过设计环节实现合作当地化。选址、地质、地势和地形特征发挥着重要作用，决定着如何营造场所[1]，如何为该场所和所在社区锦上添花。

空间被使用的时间和方式、不断变化的都市或农业景观以及整体场地战略必须留有足够的变动、扩大和调整的空间，并且适宜定居。社区的发展需要时间，以重新定位空间问题，确定空间利用方式和如何利用公共空间开展活动。充分考虑未来可能出现的种种变化以及服务和活动的多样化，这样会有助于设计出适合一系列活动的统一空间，满足不同人群的需求。

Weeksville遗产中心外立面覆盖光滑的水平木板，扫视着整个纽约布鲁克林皇冠高地街区。建筑细节和材质的选择与此地的遗产、现代都市感和社区的历史交相呼应。富有非裔美籍特色的节奏和接地气的韵律与周边的都市环境相互交织在一起，其乐融融。从水牛大道看去，景致延伸到庭院；装饰深深镶嵌在木质外立面上，从外面可以一览建筑的核

Architecture in the community acts as a bridge between the local population and their everyday activities, routes through the community, creative and learning spaces, cultural and social activities. Buildings for the community will have many roles to play, as functional and appealing architectural designs; fitting into the site and surroundings, inevitably becoming landmarks and social connectors. They may help transform the way people interact and use shared space, building better communities, offering support for local people, addressing news and events, heritage and history; a familiar place for all; a beacon, the community gateway.
How much say do the community have in the shaping of space? How does a new, shared space influence the group, the community as one?
How does an injection of new space, construction, complex urban planning, activity and event affect the individual community lives, intimately, comfortably and educationally? How do we create good, sustainable and welcoming community gateways?

Through technological innovation, original thinking, sustainability, and healthy risk taking, architecture can help create strong community identities for site and place. The architecture that represents communities must be absolutely purposeful to the local people, the community. The community must feel they understand and relate to the design and any community project will benefit from local collaborations in the design process. The considerations of site, conditions, geographic and local topographic characteristics will play an important part in the place making[1], adding the exact ingredients for that place, that community.
Time and evolution of the way space is used, the changing urban or agricultural landscape and overall site strategy must be given an element of flexibility to change, to grow and to adjust and allow for settlement. A community may need time to develop their own ways of thinking about space, using space and carrying out activities for the everyday occupancy of *shared space*. Allowing a mixed range of services, activities and anticipation for the possibilities of the future, will help create a more universal space for a spectrum of activities, for a range of user-groups.

Weeksville Heritage Center's sleek horizontal timber clad façade sweeps its statement across the site in the Crown Heights neighbourhood of Brooklyn, New York. Architectural details and materials immediately echo the heritage of the site, contemporary urban vibes and close-knit historical community chronicles. Connections to African-American beats and earthly rhythms are delightfully intertwined with the architecture in the urban setting of the area.

里斯本广场_Lisbon Square / Balonas & Menano
圣伊丽莎白东入口亭_St. Elizabeths East Gateway Pavilion / Davis Brody Bond Architecture
Weeksville遗产中心_Weeksville Heritage Center / Caples Jefferson Architects
社区通道_Community Gateway / Heidi Saarinen

心；建筑创造了阴影和反射效果，增加了遮阳处，便于人们四处走动。自然光透过玻璃长廊的正面照射进室内，整个长廊显得通透无比。建筑使用了清晰的天然材料，巧妙地应用了不同的主题，设计有空间感很强的长长的空间，为用户、活动和社区呈现出层次感。

建筑融入了创新的环境技术，例如地热温控系统，减少了化石燃料的用量。建筑用料考究，光和空间的利用充分考虑黑夜白昼的影响，提升了环境的氛围。

Weeksville遗产中心既为当地居民提供服务，也欢迎邻区和更远的参观者。这里经常举办面向当地居民和广大观众的展览，同时针对本地的实际情况举办教育活动，讲座、座谈等各项活动如期举行，图书馆和资料室等这一切让建筑充满活力，人们在此如饥似渴地获取知识，开展各项活动。

新建筑连接起一片建于19世纪的最初的非裔美籍自由民的定居地，包括发现于20世纪60年代的4座房屋。在建立社区活动的呼吁下，新建筑终于在2013年建成。

建筑师们从形状、材料入手，参考此地的传统满足了建筑要求。流动的建筑、持续的动感和条条通道带领居民穿越时空踏上寻访建筑和回忆的旅途。花园设计充满诗情画意，能唤起人们对往事的回忆和沉思，与周围环境相得益彰。生态环境和野生动物、农场的布置、简约的建筑风格、自然光和天空，只要可能，都被充分利用了起来，一切都让人印象深刻。建筑师战略性地考虑新设计，考虑保留开阔的空地、光线的流动，而不是使现存的定居地黯然失色。室内装饰淡雅、明亮，处处让人感觉积极向上。

新建筑与场地原有建筑相得益彰，并且与周边地区的历史特色相容，无论在实际环境方面，还在设计理念方面，都使新旧之间建立起相互联系。

颇具历史感的圣伊丽莎白东入口亭位于哥伦比亚区，现命名为G8大道华盛顿亭，位于72.9ha的一片土地的中央，整片土地正在进行大规模的翻新工程，准备实施一系列的计划，是城市整体规划的一部分。

随着周边重建计划的实施，混凝土和木材搭建起来的亭子成为表达友好的地标建筑，此地也举办一系列的临时活动。通过各种有趣的活动、赛事和教育活动，社区居民和此处的联系更紧密了，社区的凝聚力也增强了。

历史上，内战时此地是一处部队医院，主要治疗精神病，现今由哥

The view from Buffalo Avenue, looking into the courtyard; the deep inset slit in the timber façade, allows for peeps into the core of the building; creating shadows and reflections, adding shelter and movement. A lingering glass fronted corridor lets in natural light and transparency. Dressed in clear, skilfully themed natural materials and forms of long sweeping spaces, the architecture offers hierarchy to the occupier and activity, the community.

The building has incorporated innovative environmental technologies, such as geothermal temperature control systems, reducing the use of fossil fuels. Materials are selected with care, using light and space and considering the effects of day and night, adding ambience.

The Weeksville Heritage Center serves the immediate population and welcomes visitors from neighbouring hoods and beyond. Exhibitions for local and wider audiences are curated and local education programmes are in place; lectures, talks and events, a library and resource space all make this building come alive with diverse activity, knowledge and movement.

The new building acts as a gateway to the original 19th Century African-American Freedman's Settlement site, including four original houses that were discovered in the 1960's. An established community campaign has allowed the new gateway building to become reality in 2013.

Requirements of the brief were met by the architects through form, materiality and the smart references to the site's heritage; fluidity, continuous movement and pathways taking the inhabitant on an architectural and memorial journey through time and place. The gardens are exceptionally fitting, with poetic design moves inviting memory and mediation. Ecology and wildlife, creations of farmland and simple architectural form and preservation of light and sky where ever possible is particularly striking. The design has integrated strategic viewpoints and large openings, maintaining the flow of light, not to overshadow the existing settlement. Interiors are light, bright and a series of positive environments.

The new building fits together with the existing architecture of the site, and the historical character of the surrounding area, acknowledging physical and conceptual connections between the old and the new.

The historical site at St. Elizabeths East Gateway Pavilion, in the District of Columbia, now named the G8Way DC Pavilion, is located in the center of a 72.9-ha site, undergoing a substantial regeneration programme, covering the entire site, with a range of on-going planning proposals, part of the city's master plan.

Whilst the surrounding redevelopments are taking place, the concrete and timber pavilion structure acts as a welcoming sign post and an interim venue for a wide range of activities, strengthening the community and holding the site and local inhabitants together through the developments by hosting a lively schedule of activities, events and education programmes.

Historically, the site housed a military medical center during the Civil War; dealing primarily with mental health, and is today run by the District of Columbia, offering a vibrant vision for its local people. The pavilion acts as a catalyst for change and celebrates a

里斯本广场新人行道,广场经过翻新,可以用于举办城市公共活动
new pedestrian walkways of the Lisbon Square which was reinvented for urban and shared activities

圣伊丽莎白东入口亭既是表示欢迎的标杆,也是可以举办各种活动的临时场地
St. Elizabeths East Gateway Pavilion acts as a welcoming sign post and an interim venue for a wide range of activities

伦比亚区接管,当地居民感觉此处充满欣欣向荣的气息。亭子的建立催生了变化,让周围的居民对多彩的未来充满向往,社区建设兼收并蓄,反映了超前的理念。将整个地方建成可持续发展、社会经济一体化、兼具多功能的大熔炉是开发此地的出发点,同时要加强新建筑、当地居民、环境和现存的基础设施的相互依存关系并使其相互融合,建成可持续发展的社区通道。

地面上的开放空间摆放长形条凳,直接延伸至覆盖木板的室内空间,深受欢迎,坐在这里就可以看到全景。开放空间提供自然通风,使室内和室外衔接流畅自然。轻质的预制混凝土屋顶板和部分略高的绿色屋顶切断了水平方向的天际线,构成建筑的最顶层,形成一个完美的休憩观景台[2]。

建筑师在整体设计时充分考虑到这里要举办音乐会和表演、各种社区和教育活动。日常食品和销售自产农作物和艺术品的手工品市场,随意的就餐区和创意活动吸引许多新观众,使社区多样化,这将关乎此地和周边的今后发展。零售区出租盈利,为当地发展创造机会,同时也可以丰富居民的体验。食品卡车可以自由到达各个售货摊,便于装卸市场货物。

对这个项目而言,设计的可持续性至关重要。建筑安装了收集雨水的储水箱,风景和植物皆经过精心设计和选择,主要为抗旱品种。屋顶绿意盎然,吸引野生动物,减轻了热岛效应,也减少了运行成本。

自然通风也是设计中的一个关键因素。为鼓励人们在社区里把骑车当作一种健身方式,社区里有专门的自行车停放处。亭子是进入这片开阔空间的必经之路,属于总体规划中第一阶段的部分成果。

里斯本广场坐落于Porto市中心,是进入市区的必经入口,建筑师改变了它过去的三角形形状,重新设计了连接社区和都市住房的区域。富有纹理的混凝土外立面使新旧材料和形式里外并存,并且设计中融入了雕塑的元素。

项目的设计轻而易举地改变了一个被遗忘的城镇广场的命运,广场包括三重功用,使这片存在已久的场地变成了带给人们愉悦和具有实用性的社区林荫大道和城市公园。建筑的双重构造计划十分醒目,垂直分为两半,腾出中间的地方建成散步专用道,照明、通风、遮风挡雨遮阳设施样样俱全。日升日落、四季交替中光和影配合营造气氛,打造出不同感觉的室内景致,让人有不同的交往感受,更突显了颜色的有机搭配。

这个项目的建设取代了以前破败、存在安全隐患的场地,修建了安

diverse future for the inhabitants of the area, an eclectic and forward thinking local community. The idea is to develop the whole site into a sustainable, socio-economic, mixed-use melting pot, and to enhance and integrate the new architecture, local population and landscape closely with the existing infrastructure – creating a sustained community gateway.
The welcoming, ground level open space connects straight into the timber clad interior zones with elongated bench seating for sharing en masse. Natural ventilation through the open space creates a fluid transition between the inside and outside. Lightweight precast concrete roof panels and an elevated green roof slicing through the horizontal skyline, make up the top level of the building; becoming the perfect viewing platform for pausing and admiring the immediate locale[2].
Concerts and performance events are taking place here and use of space for various community events and educational activities has been carefully considered in the overall design programme. Regular food and craft markets, selling home made produce and artefacts, informal dining areas and creative events generate new audiences, helping the community network by adding a new level of diversity, crucial for the future of the site and surrounding developments. Retail units are let out to generate income and create local opportunities and further diversify the experiences of the local people. Food trucks have easy access to modular booths for loading and unloading market produce.
Sustainable design is fundamental to the project. Rainwater harvesting with an on-site cistern is in place, landscaping and plant species have been designed and selected with care, using drought resistant varieties. The green roof attracts wildlife and helps reduce the heat island effect, minimising running costs. Natural ventilation is key in the design. Bicycle stands encourage people to cycle as a healthy form of transport in the community. The pavilion is the welcoming entrance to this vast site, and an integral part of the first phase of the site's overall major redevelopment plans.
Set in central Porto, on a site that was once the main entrance to the city, Lisbon Square reinvents the threshold, from its triangular plan, between the community and the urban fabric. Its textured concrete façade juxtapose material and form between old and new, inside and out, with sculpturally architectural suggestions. The programme arrangement encouragingly transformed a neglected public town square, consisting of three multi purpose levels of use, giving the site a new tenacity into an enjoyable and practical community thoroughfare and urban park. The building's striking dual composition plan is divided into two verticals, giving way to a pedestrian promenade in the middle, allowing illumination, ventilation, shelter and shade. Light and shadow play with the ambience through the changing daylight and seasons, forming different qualities of interior landscapes and social experiences, enhancing the existing colour scheme.
The project has replaced what was previously a dilapidated and perilous site by introducing secure pedestrian walkways for urban and shared activities. New circulation and connection points have been created for easy movement through the site and to the city

Weeksville遗产中心，连接着原有的19世纪非裔美籍自由民的定居地
Weeksville Heritage Center, a gateway to the original 19th Century African-American Freedman's Settlement site

全的人行道，方便在城市开展大型活动。建筑师设计了新通道和连接点，便于人们出入场地内外，这些通道还通往远处的城市景观，同时也充分体现出通道自身内外的特点。里斯本广场位于城中心，吸引着散步者、购物者、大学生和途经此地前往他处的路人。

预制混凝土组件和交错的平面都说明施工方对地理条件了如指掌，具备驾驭建筑施工的能力。外立面上的白色钢材组成不规则的方格形状，逐渐延伸至现代版的略具苏格兰农场气息的地方。建筑的顶层有一个鲜活的绿色屋顶，上面四散种着橄榄树，和周边的Cordoaria花园联系在一起，让人联想到葡萄牙的Porta do Olival。在下面一层建有一个嵌入式车库，非常实用。

商业区位于中心地带，和Clerigosé塔相邻，方便欣赏Porto的美景，同时使社区的一层发挥多种功能——商店、咖啡馆和社交场所。另外，在这里还能从另一个角度观赏Clérigos塔的景致，凸显出这里的地形和历史性。

本文的话题和例举的建筑项目能引发许多有趣的讨论。不管怎么说，社区的通道影响到每个人，起到协调、疏通、激励和整合处理所有当地事情的作用，不仅社区间的联系多了，联系的范围也拓宽了。这里讨论的建筑虽然在不同的环境中，但是却有共通之处：设计面向未来，社区住着志趣相同的人，同时也欢迎前来参观的人，并且分享这片空间。换言之，被称作"社区"的地方开展的活动大致相同，甚至相同，这和当今对社区这个字眼的翻译有关。社区虽是一个普通的词，但是作为建筑师和设计师，我们会充分考虑它所蕴涵的复杂的内涵和意义。铭记这个词的含义至关重要，它能反映出社区的过去、现在和将来。社会经济和文化氛围在不同生活圈子里发生变化，但是无论人年事多高、阅历多丰富，最终都会回归家庭。

所以社区的通道一定要就位，随时欢迎新朋旧友、老者幼童、家庭或是个人。通过对本文所提项目的研究，我们了解了一系列建筑方法：建筑中对场地、遗产、叙述性的利用以及前瞻性思考后采用的技术和材料，这些因素使不同的项目呈现出不同的个性。

设计公共空间任重道远，要取悦和满足所有用户群，实现不同功能，考虑预算和美感，设计要灵活，允许后期变动，吸取过去建筑的经验教训，迎合未来的需求。

landscape beyond, equally representing the in-between; interior and exterior. Due to its central location, Lisbon Square attracts walkers, shoppers, university students and those passing through on route to nearby destinations.
Prefabricated concrete components and staggered planes hint at place and geographical and architectural control. White metal structures on the façade form irregular grid-like patterns, then gradually extend the journey towards a section of contemporary semi under crofts. The top layer of the building consists of a green living roof and scattered olive trees, creating a nexus to the nearby Cordoaria Garden, whilst referencing the Porta do Olival. Practical aspects include a built-in car park, on the lower level.
Existing commercial space in the middle section, links the Clérigos Tower, for spectacular views of Porto, giving the community a mixed-use experience on the ground level, offering shops, cafes and social areas, with sensory viewpoints from a different perspective out towards the tower of Clérigos, highlighting the immediate topographic and historical identity of this site.
The topic and architectural projects studied in this text, open up many stimulating discussions. In one way or another, community gateways affect everyone and can assist in coordinating, facilitating, inspiring and morphing together a whole range of local issues and connect local communities not just to each other, but with the wider realm. Although located in quite different geographical landscapes, the projects studied here all have something closely in common; the idea of the future for a group of likeminded people who also welcome others to visit their building, and share this space. By this I mean that the activities that take place in what we call a "community" are seen as similar; the same, even, all linking to the contemporary translation of the word community. This is a generic word and as architects and designers we take into consideration the complex content and meaning of this word. It is possible to get lost in such everyday terminology, and in words such as community. It is important to remember what the word really means, and the effects relating to community of the past, present and future. Socio-economic and cultural climates change, through various life cycles, and then return, older and wiser, to come home.
This is where the gateway to the community needs to be ready to welcome the old and new, young and old, a family or just one person. In the studies of the buildings covered in this article we see a range of approaches; usage of site, heritage, narrative and the forward thinking technology and materials all add strong identity to each one of the projects.
It is quite a task to design shared space; to please and accommodate all user groups, functions, budgets and aesthetics, to allow for change, learn from the past and through architecture; successfully celebrate the future. Heidi Saarinen

1. Cresswell T, *Place: A Short Introduction*, Blackwell Publishing Ltd., 2004, p.50.
2. Malpas, J.E, *Heidegger's Topology: Being, Place, World*, Cambridge MA: MIT Press, 2008, p.107.

里斯本广场
Balonas & Menano

里斯本广场位于波尔图市，历史上这处场地有许多增建结构。场地位于波尔图市最黄金的地带之一（场地为建筑遗产），同时也处于自1996年来的联合国教科文组织世界遗产所在地。过去，广场的前身是一个叫Anjo的市场，于1839年投入使用。市场不断地改建，以满足城市居民的需求。1952年市场被拆，但是直到1991年场地才开始重建。广场的最后面貌是面向一个室内露台的一处建有商场的空间。这种设计不久就暴露出缺陷，和周边环境不相容。11年后，这里已经丧失了它应有的功能，并且惨遭遗弃，显现出败落的迹象。

提出的方案的原则是广场或公用空间应该采用面向波尔图市的开放式外形，不再采用封闭式设计，来确保和周围的环境相互联系。5475m²的场地用于建造里斯本广场，独特的形状设计主要是要营造高度更适宜且自由的内部空间，从行人的角度去设计，以形成全新的地貌，和周边环境形成动感的关系。里斯本广场不仅是一座城中花园，同时也是一座带有一条半遮挡的商业街的建筑，建在斜坡上，三个规划层在此融为一体：花园在最高处，商业街在中间层，之前便存在的停车场位于底层。场地的上层为一座绿意盎然的广场，与Cordoaria花园和Gomes Teixeira广场毗邻。这是一处出人意料的背景，旧城的立面成为一处有趣且充满活力的场景。橄榄树凸显了建筑最高处的新花园，让人联想到之前叫做Porta do Olival（橄榄之门）的旧城门。

加强安全措施、确保留守和经过这里的人们的安全、保持公共空间的活力以及发挥它的用途是十分必要的。规划中最重要的一点是修建一条自然通道，没有障碍，通行无阻，和周围环境和谐相容。

在中间层，项目修建了一条新街道，街道上容纳了少许的功能设施，如商店、咖啡馆和餐厅。新街道犹如一条笔直的轴线，位于Clérigos塔和Lello书店之间。Lello书店是波尔图市最令人印象深刻的、拥有新艺术风格的立面和新哥特式内部设计的典范之一。对于城市中心的古街道网络来说，它是一个具有代表性的、时尚的嵌入结构。立面由带有纹理的预制混凝土和白色金属结构构件组成。

建筑师尝试寻找能够尊重周围场地遗产的解决方案，同时尝试着产生互动交流。最终场地能给小镇带来新生，提供一片开阔的空间。

Lisbon Square

This is a place that had many interventions along the time and it is located in one of the noblest areas of the city of Porto in terms of architectural heritage and in a UNESCO World Heritage area (since 1996). In the past, the square was referred to as the Anjo Market and it was inaugurated in 1839. The market suffered changes over the time in order to adapt to the needs of the city population. The last "version" of this market was demolished in 1952. Only in 1991, the site was rehabilitated. This latest version of the square was a space with a shopping arcade facing an interior patio. This solution, however, soon revealed weaknesses in the relation with the surroundings. After 11 years, the space was found disabled and abandoned, showing marks of degradation.

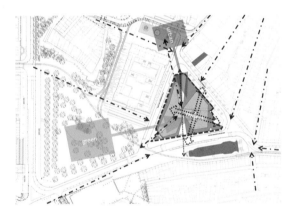

The proposed solution is based on the principle that the square/public space should adopt an open shape to the city, losing the interiority and ensuring an appealing relation with their surroundings. With 5,475m², it was proposed to the Lisbon Square, a new topography that by the singular shape driven by the need to create interior spaces with more dignified free heights, planned for pedestrians, relates dynamically with its surroundings. It is both an urban garden and a building with a semi-covered commercial street, inserted into a slope where the three programmatic levels are incorporated: Garden on the top; commercial street in the mid-level; and an already existent parking lot in lower level. For those coming from the upper part of the city, it was turned into a green square linked with Cordoaria Garden and Gomes Teixeira Square. It is a sort of unexpected setting, where the old city's facades become an interesting and lively scenario. The new garden atop the structure is punctuated by olive trees invoking one of the city's ancient gates formally known as Porta do Olival (Olival Gate). It was necessary to add safety and well-being for those who remain or pass it, ensuring the dynamics of public space and call for its use. One of the key points of the proposal was to offer a functional natural pathway, without barriers or obstacles and being concordant with the surroundings.

At a middle level, the project opens up a new street, where a few functions are located (shops, cafes, restaurants). This new street was designed as a straight axis between the Clérigos Tower and the Lello Bookstore (one of the Porto's most impressive examples of Art Nouveau facades and Neo-Gothic interiors). And it represents a fashionable addition to the rich network of historical streets in the city center. The facade is composed by a textured prefabricated concrete and white metallic structural elements.

The architect was trying to find architectural solutions which have respect for the surrounding heritage, and also attempt to communicate with. And finally it has brought a new life and an open space in the town.

项目名称：Praça de Lisboa
地点：Porto, Portugal
建筑师：Pedro Balonas, Simão Silva
助理建筑师：Pedro Almeida, Pedro Pimentel
结构、水暖、音效、电气、煤气设备、暖通系统、通信系统和安保系统：AFA CONSULT
业主：Municipality of Porto
甲方和发起人：URBACLÉRIGOS–Investimentos Imobiliários, S.A.
地产管理：JOHN NIELD&ASSOCIADOS-Gestão de Promoção Imobiliária
承包商：RODRIGUES E NÉVOA, LDA.
有效楼层面积：5,475m²
竣工时间：2012
摄影师：
Courtesy of the architect - p.100, p.104, p.108
©Carlos Azevedo(courtesy of the architect) - p.102~103, p.109
©Pedro Alves(courtesy of the architect) - p.106~107
©Pedro Brum(courtesy of the architect) - p.105, p.110, p.111(except as noted)

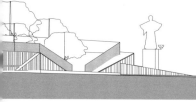

a-a' 立面图 elevation a-a'

b-b' 立面图 elevation b-b'

1 餐厅 2 咖啡厅 3 技术区 4 阳台 5 售货亭 6 商店 7 波尔图学术联盟
1. restaurant 2. cafe 3. technical area
4. balcony 5. kiosk 6. store 7. Porto Academic Federation

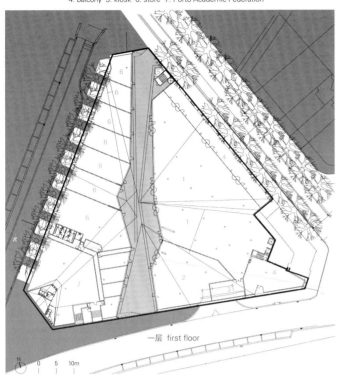

一层 first floor

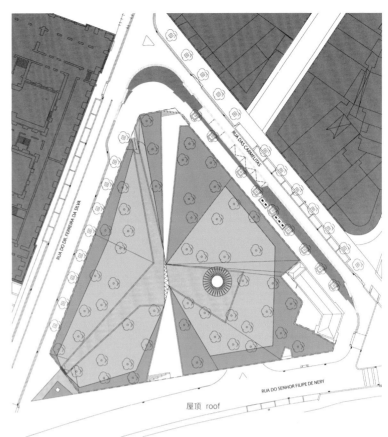

屋顶 roof

预制混凝土的应用规划图_东北立面
scheme for the application of the precast concrete_north-east elevation

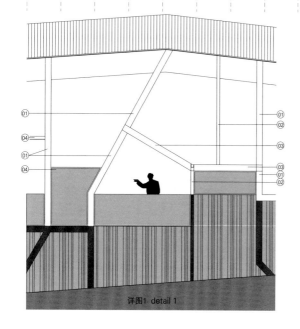

详图1 detail 1

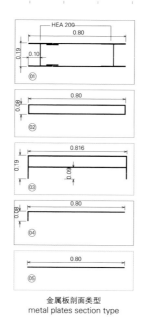

金属板剖面类型
metal plates section type

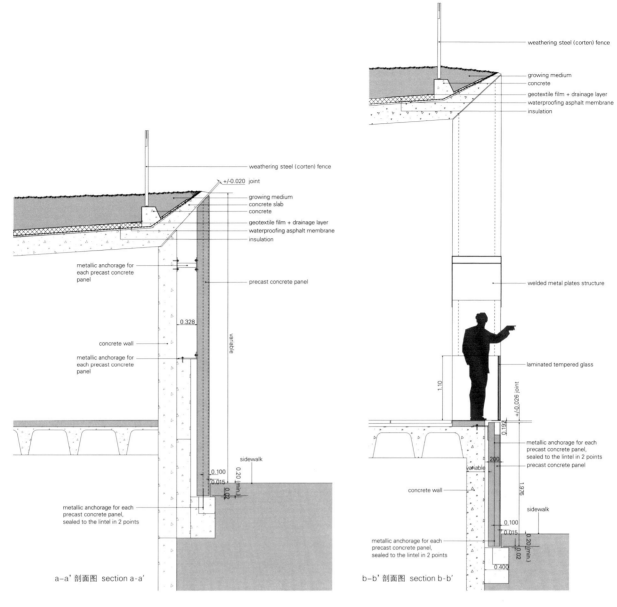

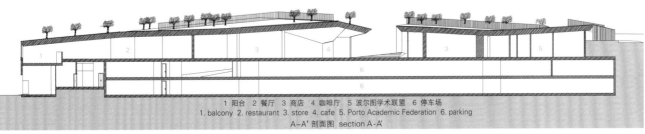

1 阳台 2 餐厅 3 商店 4 咖啡厅 5 波尔图学术联盟 6 停车场
1. balcony 2. restaurant 3. store 4. cafe 5. Porto Academic Federation 6. parking
A-A' 剖面图 section A-A'

社区通道 Community Gateway

圣伊丽莎白东入口亭
Davis Brody Bond Architecture

圣伊丽莎白东入口亭位于哥伦比亚历史街区中心第8街区内,这里有望逐渐变成一个多收入阶层入住的多功能社区。内战时这里是圣伊丽莎白医院,主要给美军提供精神疾病方面的治疗服务。现在医院大院已经不用了,联邦政府将其划拨到哥伦比亚区治下管辖。现今,约728 434m²的场地分别被公共、私人和机构投资使用。建设计划包括整合美国国土安全部和在西园区安置多达14 000名员工,圣伊丽莎白东入口亭的重建成为重要的施工项目,以实现该区发展创新型经济的目标。

圣伊丽莎白东入口亭现更名为新G8way DC,它集多种功能于一身,可以休闲就餐,有农贸市场和其他社区,可以举办文化和艺术活动。亭子在整个园区内占地约8093m²,具有很强的标志性,是易于辨认和受人欢迎的景致,尤其是能反映周边不同地区的居民原有和将来可能举办的各种活动,这一点更为有利。亭子的主要部分是一片高度约7.3m的高地,到处是模块化的摊位,便于食品货车出入此地。

亭子的设计方案是经过激烈的公开招标确定的。中标依据主要考虑如何能让建筑结构与场地完美融合在一起。"自给自足"的方法从建筑工程中获得启示,让当地人购买本地产的新鲜农产品和当地的手工品,并由各种开车贩卖食物的小贩为当地居民服务。项目建设处处体现可持续性,这点从最初的设计就可以看出来。亭子利用自备的蓄水器收集雨水,满足灌溉用水之需;景观设计专门种植抗旱植物;屋顶种植的植物可以减少热岛效应,减少了这片封闭区域对机械设备系统的要求。而机械设备系统严重依赖自然通风。这个概念不仅考虑到它理想的实用性,同时使用起来灵活自由不受限。双层设计便于人们在不同层面行动和活动。地面一层的设计有利于从各个重要的边缘地带进出此地。一层清晰地一分为三,将建筑的都市感和绿草盈盈的景观结合在一起。屋顶层可供行人自由行走,可以从全新的观景视角欣赏周围的街区。种植拓展型植被的高架屋顶适宜开展多种活动,包括午后音乐会和社区活动。

以前这里仅允许少数人自由出入,G8way DC的新设计使所有人可以进出,欢迎大家前来同聚屋檐下,品尝美食、学习和娱乐。

St. Elizabeths East Gateway Pavilion

Located at the center of the District of Columbia's historic neighborhoods in Ward 8, St. Elizabeths East is on its way to becoming a viable mixed-use, mixed-income community. Once the site of St. Elizabeths Hospital which provided mental health care to the US military dating back to the Civil War, the campus has been decommissioned and transferred from the federal government to the District of Columbia. Today the 180-acre site provides a setting for public, private, and institutional investment. With the planned consolidation of the Department of Homeland Security (DHS) and location of over 14,000 employees on the West Campus, the redevelopment of St Elizabeths East has become a critical project in realizing the District's goals of cultivating an innovation-based economy.
The new G8way DC, formerly known as St. Elizabeths East Gateway Pavilion, is a multi-purpose structure providing a venue for casual dining, a farmers' market and other community, cultural and arts events. The Pavilion, spread over a two-acre plot of the campus, creates an instantly iconic, visible and welcoming view into the

项目名称：Saint Elizabeths East Gateway Pavilion – G8WAY DC　地点：Washington, DC
建筑师：Davis Brody Bond Architecture
设计团队：总负责人_Peter Cook/项目顾问_Will Paxson/项目设计师_Rob Anderson/Cody McNeal, Ryan Meyer, Adam Grosshans
结构工程师：Robert Silman Associates　土木工程师：A. Morton Thomas & Associates　电气工程师、可持续性、照明：WSP Flack + Kurtz
景观建筑师：Gustafson Guthrie Nichol　总承包商：KADCON Corporation
甲方：The District of Columbia Department of General Services
摄影师：©Eric Taylor (courtesy of the architect)

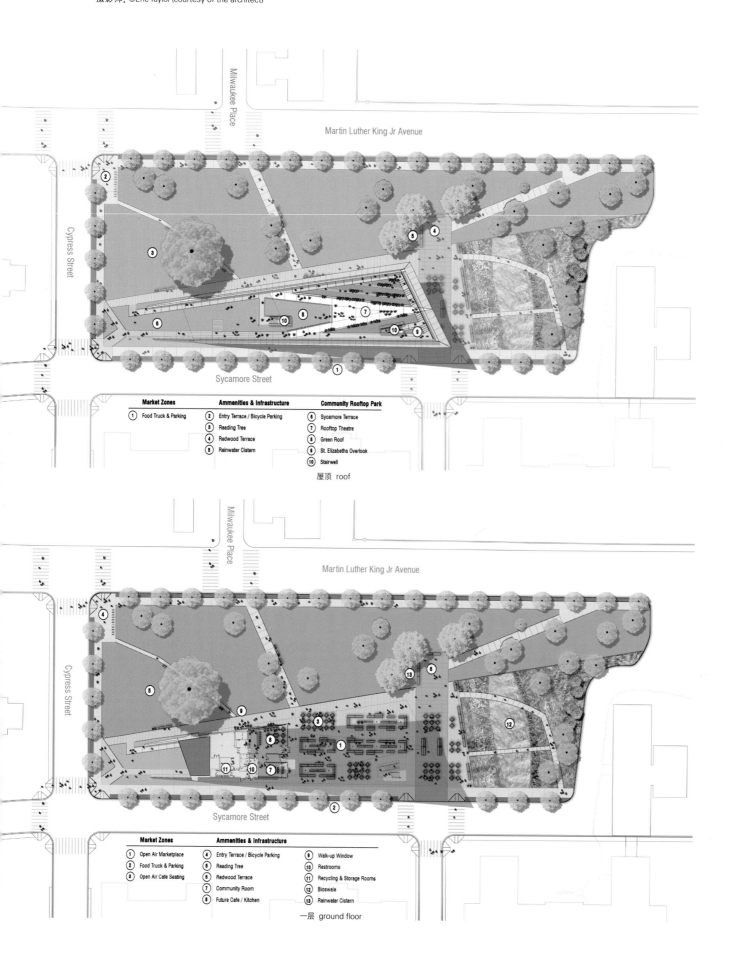

site, particularly from the vantage points that reflect the existing and anticipated movements of people from different areas of the neighborhood. Forming a dramatic backdrop to the plaza, the main area of the pavilion is a 24-foot high space filled with modular booths convenient to where food trucks access the site.

Selected in a highly publicized design competition, the pavilion focuses on the seamless integration of the structure and the land. This "of the land" approach takes its cues from the architectural program which is centered on the provision of fresh produce, locally made crafts and a variety of food truck vendors to serve the local population. Sustainability is integral to the project, informing the design from the outset. The pavilion employs rainwater harvesting (i.e. an on-site cistern captures runoff and supplies the entire site's irrigation demand); the landscape design provides for drought resistant plantings; the roof plantings reduce the heat island effect which reduces demand on mechanical systems in the enclosed portion; and, the mechanical systems rely heavily on natural ventilation.

The concept incorporates the desired functionality while at the same time providing for flexibility and spontaneity. The dual-level design allows for movement and activities throughout the site at different levels. The ground level encourages easy connections from the most prominent edges of the site, creating three distinct zones, and connecting the urban face of the project to the more pastoral campus setting. The roof level access allows pedestrians to gain a new perspective on the neighborhood by moving seamlessly up and across the site along the universally accessible roof level. This elevated landscape includes an intensive green roof where multiple activities can occur, including afternoon concerts and community events.

Once inaccessible to all but a few, the design of G8way DC will open the site to all, welcoming people to come together under one roof to eat, to learn and to have fun.

社区通道 Community Gateway

Weeksville遗产中心
Caples Jefferson Architects

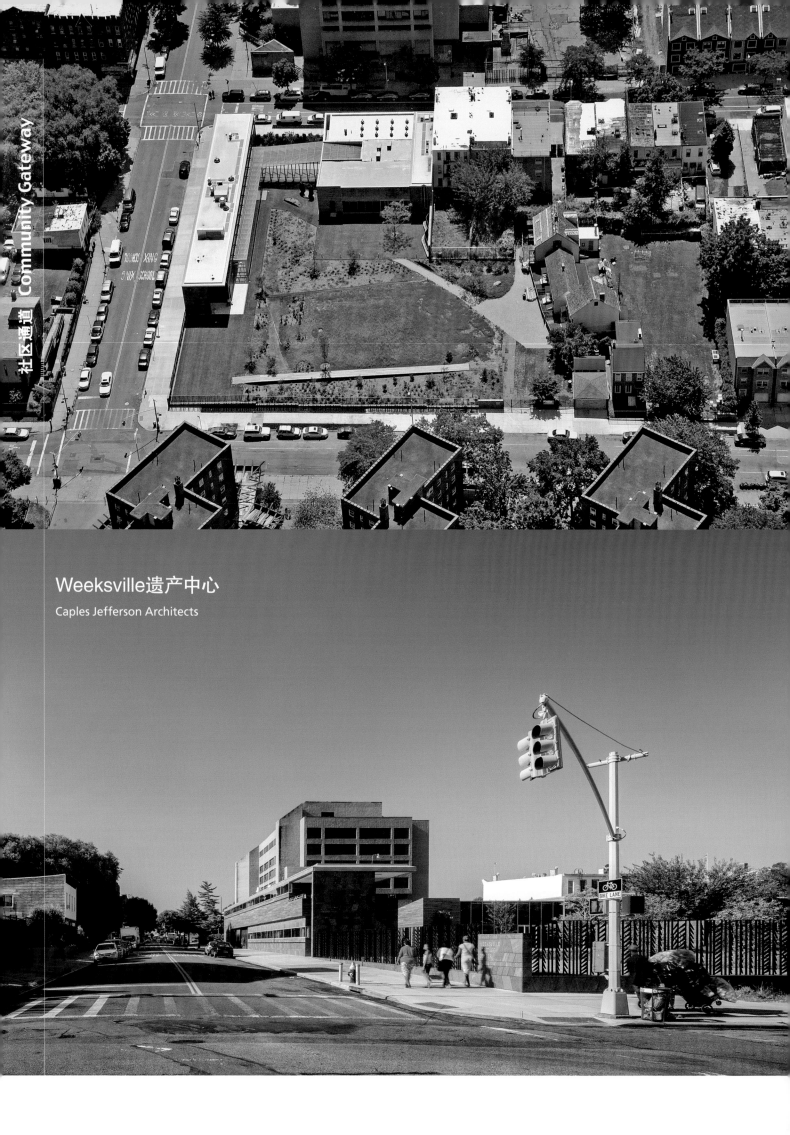

Weeksville遗产中心设计独特，具有现代建筑的简约风格，条形窗和玻璃走廊处处体现出非洲元素。多种多样的元素形成了视觉反差。这些重复的元素整合在施工和结构构件中，例如在铺路材料和石头的接缝处、颜色和材料的选择、围栏柱子的细节和遮阳玻璃上的釉料中。虽然现代建筑语汇和非洲元素自成一体，但是总体来看还是很和谐的。

新建筑位于Weeksville遗产中心这片具有历史意义、常唤起人们缅怀之情的地方，此时此地全新的建筑外观重新诠释Weeksville所在地的背景。

Weeksville遗产中心是可持续利用的现代建筑，连接起19世纪非洲裔美国人被解放的奴隶的定居地。19世纪60年代，在这块布鲁克林定居地上发现了四座遗留的建筑。

在社区40年间的大力支持下，这里最终变成了非洲裔美国人的遗产核心地。Weeksville遗产中心标志着脱离过往那段历史后新生的宣言。

新建筑和景观设计的首要目的是建立一条途径，使人们了解建在此地的具有历史意义的住宅，主要是19世纪Weeksville自由的非洲裔美国人社区遗留下来的，带有先进的艺术展览、表演和教育设施，同时也给参观者和当地社区提供一片绿洲。主厅陈列介绍性展品，经由大厅可通往观赏不同展示品的画廊，继续往前有一处可容200人的讲座和表演空间，也有供参观小组和社区开展教育活动的教室。图书馆资料可供来访学者使用。管理办公室设在二层，地下室用来存放档案，另有一间屋子专门用于归置口述历史档案。

景观是这片土地规划的亮点。这处空间作为过渡，位于古宅和新中心的中间。穿行于人为修整的农田间犹如从现在走向过去，此时与彼时不分。

绵延的草地和大片的野花让人想起这个社区以前是靠农业为生。

从旧地图推论得知，"猎人飞道"这条旧道在"魅影景观"的房屋前时隐时现。

出于对历史古宅的敬重，建筑有意建低，选址也是出于对老房子的保护。建筑同时建有一扇开敞的大门，沿旧的印第安小径可直达房子，也可经由透明的走廊看到宽阔的长长的历史遗址。

建筑的外围护结构包括一个木质挡雨屏、石板瓦挡雨屏、中空玻璃窗组成的一面墙和水平的条形窗。木质挡雨屏由特制的铣削重蚁木组成，带有开放的接缝，连接在铝夹上，并带有连续的气密层。石板瓦挡雨屏由约3cm厚的定制石板构成，借助石质锚固件安装在金属承重立筋上，并带有连续的气密层。屋顶上覆盖的中空夹胶玻璃薄片上有专门设计的彩釉图案，和非洲风格图案相呼应，同时起到遮阳作用。非洲装饰和雕塑的元素在建筑中有意识地重复运用。阳光突出了建筑的主题：重复、运动、韵律和比例。这些构件在阳光的照射下显示出丰富的光影互动效果，使它们具有不同的氛围和多变的外观。

Weeksville遗产中心和当地社区保持紧密联系，这其中包括附近住户达2400人的金斯布罗公共住宅开发项目。夏天的时候，weeksville在每周六变成了农贸市场，有时还举办免费的夏季音乐会并连续放映电影。新建筑包括约3716m²的开放景观供社区使用。项目所在地和好几条市区公交车线路相邻，走路10分钟可到达3条地铁线，离最近的火车站走路仅用20分钟。

Weeksville Heritage Center

Weeksville Heritage Center is a unique urban design project in which a modern architectural syntax of simple forms, strip windows, and glass passageways is impacted by repeated African riffs. The riffs are variations that provide a visual counterpoint. The riffs are embedded in construction, in structural elements, such as the joints in paving and stone, in the choice of colors and materials, and in the details like the fence posts and the frit in the sunshading glass. The modern syntax and African riffs, although independent from each other, harmonize when experienced as an entirety.

Placed within the historic site of Weeksville Heritage Center and its evocative landscape, the new work redefines the context of Weeksville at this specific time and place.

Weeksville Heritage Center is a sustainable modern building that serves as the gateway to a 19th Century African-American Freedman's Settlement. In the 1960's, 4 remaining buildings from this Brooklyn settlement were rediscovered. Through 40 years of impassioned community support, the site eventually grew to serve as a focal African-American heritage site. Weeksville Heritage Center is the latest manifestation of that history coming alive.

The primary purpose of the new structure and landscape is to serve as a gateway to the historic houses on the premises – remnants of the 19th century free African American community of Weeksville – with state of the art exhibition, performance and educational facilities, as well as to provide a green oasis for visitors

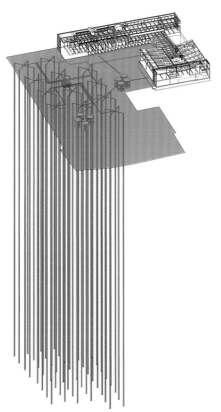

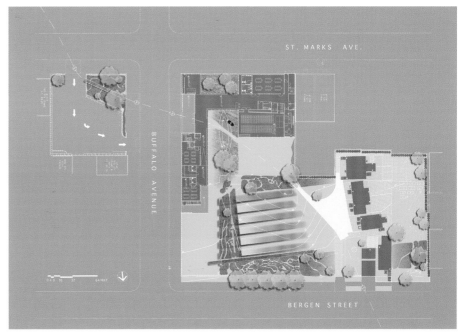

and the local community. The main lobby will include introductory exhibits, and leads to a gallery for changing shows, a lecture and performance space for 200, classrooms for visiting groups and for community education, and a library resource center for visiting scholars. Administrative offices are to be located on the second floor, and the cellar is to include archival storage space as well as a room for recording oral histories.

The landscape is the dominant element in the composition. This space creates a transitional distance between the historic houses and new center. Movement through the recreated farmland links the present to the past, between the now and the then.

The rolling mown field, and areas of wildflowers evoke the community's agricultural origins.

The old trail "Hunterfly Road" disappears and reappears before the houses in a "ghost landscape" extrapolated from old maps.

In deference to the historic structures, the building is kept intentionally low, sited to protect the view of the old houses, while providing the broad portal gateway along the old Indian trail to the houses and long open views of the historic site through the transparent corridors.

The building enclosure consists of a composition of wood rainscreen, slate rainscreen, and insulated glass window walls and horizontal ribbon windows. The wood rainscreen consists of specially milled îpe boards, with open joints, attached to aluminum clips over a continuous air barrier. The slate rainscreen consists of 1-1/4" thick custom-cut slate panels mechanically attached to load-bearing metal studs with stone anchors, over a continuous air barrier. The laminated insulated glass roof includes a specially designed frit pattern, echoing African patterns, for solar shading. The subliminal perceptual stimuli decoration and sculpture unique to Africa, the architectural ideas of repetition, movement, rhythm, and proportion are revealed by sunlight. Natural light enriches these riffs by adding shadows, moods, and ever-changing perspectives.

The Weeksville Heritage Center organization maintains deep ties to the local community, including the 2400 residents of the neighboring Kingsborough Houses public housing development. During summer months, Weeksville hosts community farmers markets every Saturday and stages a free summer concert and film series. The new building includes a 40,000 square feet open landscaped area for community use. The project site is immediately adjacent to several municipal bus lines, a 10-minute walk to three subway lines, and a 20-minute walk to the nearest regional rail station.

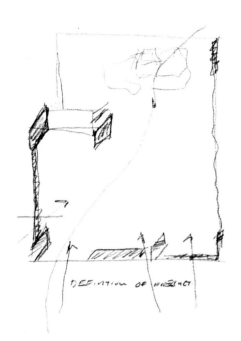

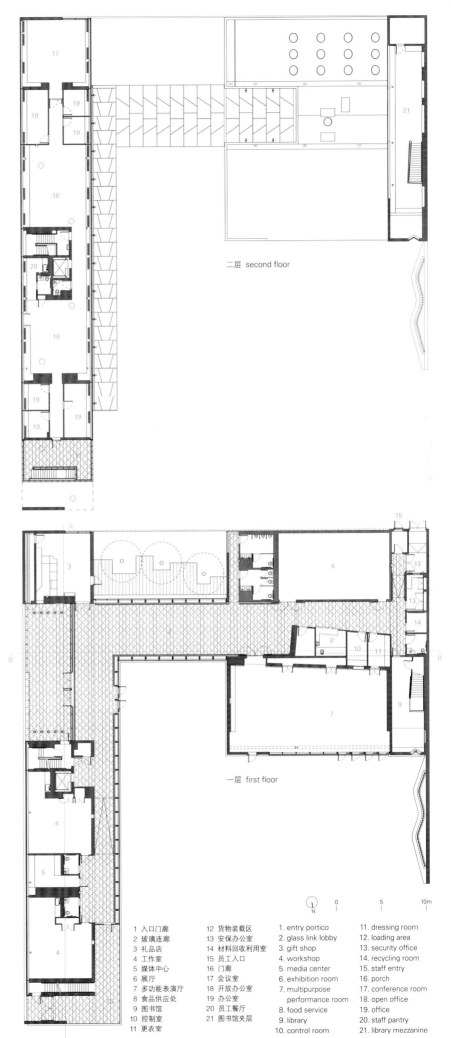

二层 second floor

一层 first floor

项目名称：Weeksville Heritage Center
地点：Brooklyn, New York, United States
建筑师：Caples Jefferson Architects
施工经理：Hill International
总承包商：Brickens Construction
结构工程师：Severud Associates
电气工程师：Loring Consulting Engineers
土木和岩土工程师：Langan Engineering
地热工程师：.W. Grosser Consulting
照明设计：Berg-Howland Associates
景观建筑师：Elizabeth Kennedy Landscape Architects
成本预算：Faithful + Gould
声学与视听：Shen Milsom + Wilke
剧院照明：Cline Bettridge Bernstein Lighting Design
建筑部门：Metropolis
供货商：Heller + Metzger PC
幕墙：Gordon Smith Corporation
可持续性设计与运行：Viridian
安保：Ducibella Venter & Santore
博物馆规划：Dial Associates
甲方：David Burney, Victor Metoyer, Pamela Green
场地面积：3,800m² 建筑面积：1,370m²
总楼面面积：2,140m²
设计时间：2008 竣工时间：2013.8
摄影师：©Nic Lehoux(courtesy of the architect)

1 入口门廊	12 货物装载区	1. entry portico	11. dressing room
2 玻璃连廊	13 安保办公室	2. glass link lobby	12. loading area
3 礼品店	14 材料回收利用室	3. gift shop	13. security office
4 工作室	15 员工入口	4. workshop	14. recycling room
5 媒体中心	16 门廊	5. media center	15. staff entry
6 展厅	17 会议室	6. exhibition room	16. porch
7 多功能表演厅	18 开放办公室	7. multipurpose performance room	17. conference room
8 食品供应处	19 办公室	8. food service	18. open office
9 图书馆	20 员工餐厅	9. library	19. office
10 控制室	21 图书馆夹层	10. control room	20. staff pantry
11 更衣室			21. library mezzanine

西立面 west elevation

北立面 north elevation

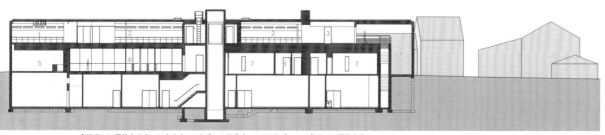

1 会议室 2 开放办公室 3 办公室 4 门廊 5 礼品店 6 入口门廊 7 工作室 8 媒体中心
1. conference room 2. open office 3. office 4. porch 5. gift shop 6. entry portico 7. workshop 8. media center
A – A' 剖面图 section A-A'

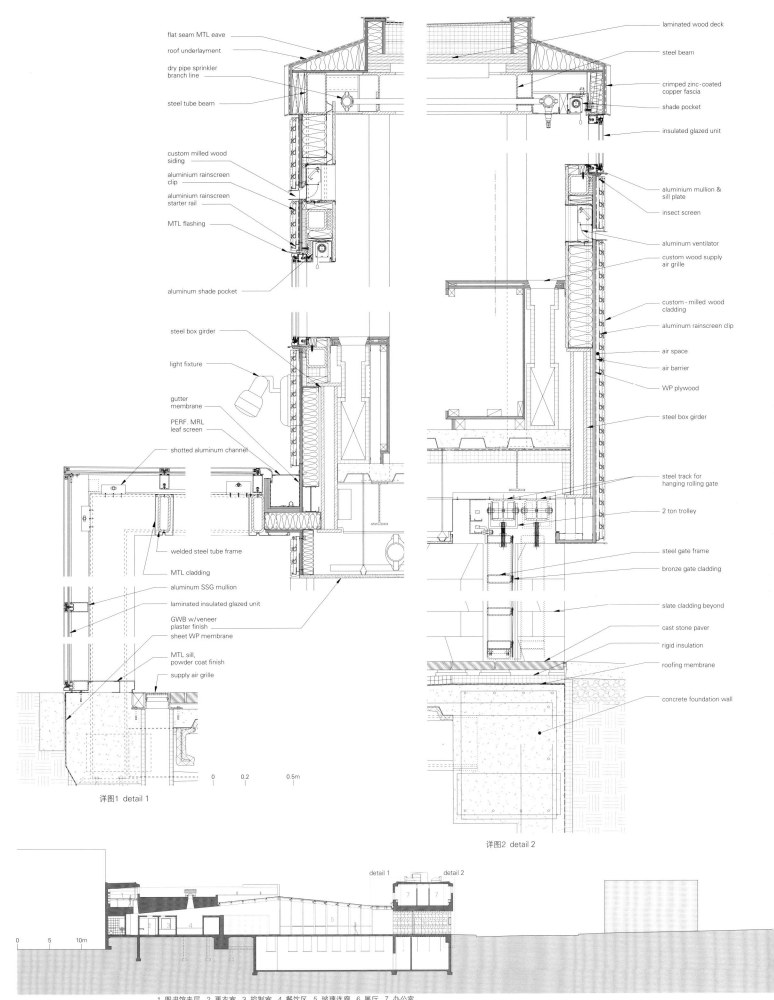

艺术高度

Artificially High

前面讨论的主题是如何采用不同的方法使建筑和建筑所在地有机地融为一体，这里我们呈现给大家的建筑项目采用的建筑方法和上述方法截然不同。建筑尝试离开地面，极尽所能地表现得轻巧，因为轻巧所以可以矗立在风景中，占据新高，开阔视野。

像《原子的达利》这一摄影作品一样，这些项目突破限制，追求悬空感，建筑材质尽量选轻的，和陡峭崎岖的地形连在一起。勒·柯布西耶和密斯·凡·德·罗恰如其分地帮我们界定出两种截然不同的方式实现建筑的轻巧，虽然我们追求设计能从周围的风景中获益，但是我们亦希望建筑和这片土地能相得益彰。

After the previous issue, in which was described a method of approach to difficult sites defined by a strong thrust towards integration with the territory, here will be presented projects in which research takes rather the opposite approach. It attempts to leave the ground, striving for extreme levity, a lightness that allows it to stand upon the landscape, to construct new heights, to open new views.

As in the work Dalì Atomicus, these built projects push themselves toward a moment of suspension, in which masses shed their weight and manage to cling to steep, difficult terrain: The examples of Le Corbusier and Mies Van Der Rohe will help us define two distinct approaches to lightness, describing worlds in which, although we seek a great opening to the surrounding landscape, we return again to a strong integration with the territory.

达利的轻巧设计

最终的姿势是经过28次尝试最终确定下来的：三只飞翔的猫、呈S形的喷水框架、一把椅子和两幅油画悬挂在房中，当然还有达利跳跃并手持调色盘和画笔的画面。

这是菲利普·哈尔斯曼1948年的摄影作品《原子的达利》。

照片表现出惊人的轻盈、悬空的愉悦感，同时也融入了西班牙艺术家的超现实主义视角。多种组合（飞翔的猫、背景中变形的水、达利的跳跃）足以震撼我们的心灵，让我们屏住呼吸，等待每件事物遵从重力原理恢复原形。总之这一切呈现出完全不同的故事。

但是，万物皆在，晶莹透亮地以固体存在，沉浸在梦中的世界里，却又无法苏醒过来。虽然有光，但拍摄照片的过程中光的出现绝对不是偶然：要验证艺术家创作的证据，即照片诞生的原因，意味着我们必须把自己沉浸在并行的世界中，我们所看到的作品中表现出的轻盈是经过漫

A very Light Version of Dalì

It took twenty-eight attempts to find the correct pose: three flying cats, a frame-spanning spray of water bent into the form of an S, a chair and two paintings floating in the room and, of course, Salvador Dalì, jumping while holding a palette and brush.

It is Philippe Halsman's 1948 photograph Dalì Atomicus.

This picture evinces a surprising lightness, a sense of joyful suspension, filtered by the surrealist vision of the Spanish artist. The combination of events (the cats flying, the water deformed within the setting, Dalì's jump) is truly able to reproduce a state of mind that makes us hold our breath, waiting for everything to return to a normal state, complying with the laws of gravity, that, in general, tell a different story.

Yet everything is there, crystallized in consolidated form, enrapt in a dream world from which we cannot, fortunately, wake up.

Light though it may appear, the photograph was by no means, however, lightly fortuitous in the making: To study the artist's proof and thus the photograph's genesis means to immerse oneself in a parallel world, in which all the lightness of the final version which we feel seeing the work becomes an arduous journey of

图拉住宅_Tula House/Patkau Architects
Solo住宅_Solo House/Pezo von Ellrichshausen
悬崖上的住宅_House on the Cliff/Fran Silvestre Arquitectos
车库房_Car Park House/Anonymous Architects
石屋_The Pierre House/Olson Kundig Architects

艺术高度_Artificially High/Diego Terna

长的、无数次的尝试和失败才获得的，充满小小的不为人知的技巧和构图的简约。总之，它们让我们意识到要臻于完美或提升作品的高度需耗费时间方能实现。

可能唯有如此，我们才能真正和意象通灵，展现出它无上的轻巧，唯有付出努力才能实现。

长腿生眼

在之前的书中，西蒙·考达谈到了利用空间法实现建筑与自然环境的融合，善待建筑周边的环境，因为建筑自身的流动性会自然地融入到地域的不同形式当中。

但是，我们将会看到有些建筑在特定背景下呈现出的形式会挑战地心引力，试图挣脱环境的束缚，这种做法与其说是对自然的不尊重，不如说是人在巧妙地和自然对抗。

我们可以看出这里对待外部环境有两种截然不同的态度，但是都追求悬空与轻盈。

为了达到这个目的，建筑物首先要有支柱（腿），将建筑从地面上抬起，有时地面是不规则的。

这样的设计使我们想到了勒·柯布西耶，1926年他提出了《建筑五要素》观点，第一点，可能也是最重要的一点，至少也是这条定义中最具修改性的一条是新建筑存在的根本是柱基。

这个理论并不陌生，千年来，人类一直在建造房屋，如果有必要，房屋会用柱子托起，建筑和地面之间会有足够多的空隙，避免洪水侵袭、食肉动物伤人，同时观景视野更好。

这位瑞士建筑师的用意是采用新的方法来解释古代建筑的空间机制，完全依赖环境、自然和建筑三者之间的关系来实现，即建筑变成了世界舞台上的演员，充满无限魅力，并且只要可能，尽量不接触周边的

failed attempts, of small hidden tricks, of rough errors. They show us, in short, a time-consuming striving towards perfection, or, rather, the work's heightened artificiality.
And, perhaps, in just this way we can arrive at absolute empathy with the image, which simultaneously presents the most extreme lightness and the heavy effort it took to produce that lightness.

Building some Legs, Providing Eyes
In the previous issue, Simon Corda spoke of a spatial approach that seeks integration with the natural topography, an attitude that respects the environment surrounding the architecture, as the architecture itself flows, naturally, into the forms of the territory.
Here, however, we will see projects that find in the forms of the context a challenge to gravity, an attempt to break away from the environment, an attitude not of disrespect but, at most, of confrontation between artifice and nature.
Here we can recognize two attitudes towards the external environment, both speaking of a search towards upwardness, towards levity.

To that end, the building can first of all acquire legs, which raise it from the ground and settle it onto the surface, sometimes irregular, of the soil.
Such is the lesson that brings us, in 1926, to Le Corbusier, when he advocates *the five principles of a new architecture* and, in particular, when he speaks of the first point, perhaps the most important or, at least, the one most amenable to definition, the genesis of a new architecture: the *pilotis*.
This principle is clearly not new: for millennia humankind has made buildings, when necessary, that are supported on stilts, which allow a functional gap between structure and ground, avoiding flooding, providing protection from predators, commanding a better view.
Yet there is in the intentions of the Swiss architect a new way of understanding the ancient spatial mechanism, which exists entirely within the new relationship that is established between the environment, nature and architecture: The building, that is, becomes an actor playing on the stage of the world with infinite grace, touching surrounding nature only at very limited points, in an expression of extreme respect for that nature, by admitting

《原子的达利》,萨尔瓦多·达利的肖像摄影作品,菲利普·哈尔斯曼于1948年拍摄
Dalí Atomicus, Salvador Dalí's portrait photo, by Philippe Halsman, 1948

环境,这是出于对那里自然的绝对尊重,意识到建筑物下小环境的存在,但不破坏土层,不侵占土地。

如此一来,我们必须接受拟态建筑依然建在地面上,建筑构造依然不变的事实,这并不总是生态环保,或是尊重自然;这么看来,有柱基的建筑可能因为更容易被看见,导致在这片风景中会受到影响,但事实是,从某种程度上讲,它有着更好的生态足迹。

地形复杂尤其如此,比如下面要谈及的项目。

靠底层架空柱抬高的楼层获得的轻盈感得益于它奋力离开地面的努力,可以通过对风景加以技巧性的利用实现,正如哈斯曼千般努力的完美拍摄一样成就了有风景的房间。

在勒·柯布西耶提出《建筑五要素》的同一时期,密斯·凡·德·罗的杰作布尔诺图根达别墅也完美竣工。

选址的斜坡问题带来的困难通过减小通风处的方法解决了,地面支撑起沉重的结构,打破常规,将入口设在上面(一般由下而上登堂入室,而这个项目是从上往下进入室内)。简言之,这样的做法和前一种模式的轻盈正好相反。

进入房子后沉重感瞬间消失,房子变成了一个飞翔在风景上的物体:客厅通过长长的玻璃窗向外延伸,在客厅会有朝下降落要拥抱整片地域的感觉。

就好像去除了外加的遮挡物,自然和周围的环境突然映入我们的视线,之后很长的时间,因为墙不透明,我们只能看到几英尺远的地方。

在这里,建筑无限制地朝周边延伸,好似一只睁大的眼睛望向外面。如果按照勒·柯布西耶的想法,建筑会呈现垂直的姿势,但密斯·凡·德·罗对物体的形状不感兴趣。

原子建筑

因为这些人的态度,从这些建筑中可以看到寻求极度轻巧的迹象,我们认为这种因素在日常生活中也非常重要。

如果依照严格的、不可变通的模式来给上述案例分类,那是没办法分类的,与其这样,不如说上述分析的建筑与周边的环境保持不断的交流对话。

Anonymous建筑师事务所建造的车库房,事实上起到中和这两种形

the surroundings under the building, without upsetting the soil, without consuming territory.

In this sense, then, we must accept the fact that the mimicry of a building that is still on the ground and follows its conformation, it is not always synonymous of ecology, or of respect; in this sense, the buildings that stand on their legs could possibly be affected by a greater visibility in the landscape, but in fact they have, to some extent , a better ecological footprint.

Such is the case even more so when the topography is difficult, as is the case for the projects presented below.

This sense of lightness that a pilotis-raised floor achieves based on its striving to leave the ground can be obtained via scenic artifice in the spirit of Halsman's efforts to obtain the perfect shot: the *room with a view*.

In the same period when Le Corbusier theorized the five points of architecture, Mies Van Der Rohe brought to fruition an absolute masterpiece: *Villa Tugendhat* in Brno.

Here the difficulties emerging from a sloping site are addressed using a less airy approach, anchoring the structure to the ground with heaviness, employing an entrance from above, and thus subverting normal plan relationships (instead of going up into the house, we descend): It evinces an attitude, in short, that seems quite far from the lightness of the first model.

Yet, entering the house, one arrives in a place that suddenly eliminates any heaviness, turning the structure into an object flying over the landscape: We are in the living room, opened to the exterior via a long glass window through which we fall, embracing the whole territory.

As if eliminating an imposed blindness, we suddenly gain a view of nature, of the surroundings, across vast sightlines, after which, for long moments, we see only a few inches from the opaque walls.

Here the building, which follows the intent of Le Corbusier would acquire an upright posture, and Mies Van Der Rohe loses interest in its physical conformation, as it stretches fully towards the surrounding landscape, a large eye intent upon the outside.

Architecture Atomicus

Based on these attitudes, one may clearly identify in the presented projects an extreme search for lightness, which we recognize as an element of great importance in everyday life.

It is not possible to divide these cases according to steely, immutable models; rather, the analyzed buildings live in constant dialogue with their surrounding landscape.

The Car Park House by Anonymous Architects establishes, in fact, a connection between the two models, but emphasizes their primary features: The pilotis become a massive septum and the house acquires a heaviness, such that the opening to the land-

捷克共和国布尔诺图根达别墅,密斯·凡·德·罗于1930年设计。独立支撑的三层住宅建在斜坡上
Tugendhat Villa in Brno, Czech Republic by Mies van der Rohe, 1930. The free-standing three-story residence built on a slope.

式的作用,但又强调了它们的主要特征:地面架空柱变成一个巨大的中隔,整座房屋显得沉重,这样一来,风景的入口处实际上是在这个相同的体量上切割出来的。车成了这座房子空间的主角。如果换作密斯来设计的话,房子入口仍然会用来平衡公共街道和人体尺度,但是现在这里只有街道,建造的房子只需考虑发挥车子的作用问题。

甚至帕特考建筑师事务所设计的图拉建筑也兼收并蓄两种方法,只是建筑的亮点是围绕周边美丽的风景来突出自己,所以使建筑脱离地面的柱子变形成最迷人的悬臂梁,朝向外面延伸。虽然施工是围着中间的天井进行的,但是整体构造却倾向有水的一边。整座建筑坐落在不稳当的岩石上,但这却成了最吸引房子住户的别开生面的场景。

悬崖上的住宅由费朗·休威特建筑事务所设计建造,主题是改变建筑形状来界定风景。

费朗·休威特建造的房子努力朝上伸展,犹如在这片风景上空建了一个窗户,设计努力要赋予建筑一种飞翔的兴奋感,让人产生一种视觉印象,感觉建筑是开阔风景的组成部分,就要从岩石上落入下面的海里。白色的建筑是现代主义建筑师耶·柯布西耶和密斯·凡·德罗的典型风格,但在这里变成了界定抽象空间的一种方式,一种视觉工具。

从某种程度上来说,奥尔森·昆丁建筑事务所建造的石屋也尝试从地面上站起来,克服艰难寻找光和景色。同样,这座建筑也发生了形式变化,我们似乎可以看出密斯的风格,不同的是建筑更倾向朝向外面开放,尝试把窗户当双筒望远镜而不是空中平台。

Pezo von Ellrichshausen建筑师事务所设计建造的独立住宅矗立于风景中,颇有帕拉奥的圆形别墅的风范:四面相同的外墙、中间有天井,几何的纯粹和天然的周边环境融为一体,这些因素明显和这位意大利建筑师的杰作相互影响。

比起其他项目,这里介绍的建筑项目更明晰地显示出盘旋于风景之上的能力,仅一根粗大的混凝土柱子足以支撑起建筑的内部空间。

楼板之上的房子像是飞过森林,框出新的景色,与自然景色完全融入一体。

现在回头再看《原子的达利》,那时我们可能感觉所有的东西都会掉下来,但是这并没有发生,因为我们能够定格那个场面,无限,轻盈,直至画面之外。

scape is actually a cut in the same volume. The car becomes a pivotal character in the spatial story of the house: If in Mies the entrance to the house was still a mediation between the public street and the human dimension, here there is nothing but the street, and the house seems to be built toward the function of the vehicle.

Thus, even the Tula House by Patkau Architects pursues a mix of the two approaches, though it seems the architecture defines itself around the stunning views outside, and therefore the same pillars that facilitate detachment from the ground, deform themselves into the most striking cantilever towards the view. Even as it rotates around a central patio, the overall composition is biased towards the water, sitting unstable on the rock – an almost dramatic scene that captivates the house's inhabitants.

House on the Cliff by Fran Silvestre Arquitectos works on a theme of deforming masses to define views.

Fran Silvestre's House is stretched entirely upward to build a window on the landscape: The structural effort is clearly designed to impart the sense of the exhilaration of flight, offering a visual suggestion of being part of a broader landscape that shatters from the rocks into the sea below. The structure's whiteness, typical of the modernist architecture of Le Corbusier and Mies van der Rohe, here becomes a means of defining an abstract space, transforming it into a device devoted to the function of seeing.

In a way, one sees even in the Pierre House by Olson Kundig Architects, an attempt to start from the existing in order to stand up, thus surmounting roughness to find light and views. Here too, the structure is deformed and we observe references to the Mies house, but with a pronounced overture to the outside, a research with windows as binoculars rather than flying platforms.

The Solo House by Pezo von Ellrichshausen Architects is a device that stands in the landscape with a clear reference to Palladio's Villa La Rotonda: The four identical facades, the central patio, the geometric purity that interacts with the organic nature of the surroundings are obvious elements that interact directly with the masterpiece of the Italian Architect.

Yet here, even more so than in the other projects, is clearly stated the ability to hover above the landscape, through what becomes a single large concrete pillar, sufficiently large to acquire its own internal space.

It is over this floor that the house can fly over the forest, frame new views, dialogue with nature as an element absolutely integrated into the landscape.

And so we come back to *Dali Atomicus*, at that moment at which we feel the obligation to fall, but the fall does not happen, because we are able to freeze a situation, infinity, light, toward the outside. Diego Terna

住宅 艺术高度 Dwell How Artificially High

图拉住宅
Patkau Architects

图拉住宅位于高出太平洋13.4m的一个偏远的小岛上,住宅的设计处处显出选址地岩石矿脉的不规则、海滩和森林的几何和空间规律性。

住宅所在地地形不规则,景致也是多样的。东边的景致包括开阔的水域和乔治亚海峡上的岛屿,绵延至不列颠哥伦比亚省内陆上的山脉上。南边俯视一个不大的潮汐盆地。广阔的绿树林和长满植被的岩石裂缝中、峡谷和洼地中随处可见玄武岩山,上面长满苔藓。一排排的赤杨和阔叶枫树使黑黝黝的道格拉斯冷杉林变得生机勃勃。住宅下面的海岸线上随处可见丢弃在海上的垃圾,原木和岩石像孩童的玩具一般被潮水投来掷去。可谓一地多景。

住宅建造时正好利用了此地的多样性。砾石铺就的小路从低矮的岩石墙边开始,一直延伸到住宅处。水泥墙随处可见,水泥墙外包裹着交错排列的黑色纤维水泥板,营造出空间感。从远处看,住宅犹如隐没在黑森林里。屋顶上种着苔藓和在当地生长的植被,从上往下看感觉是地平面的延续。

地下水缓缓流经此地,在入口的院子处被暂时截流,院子的地平面和住宅的内部地面上铺着大块碎玻璃似的水泥板。入得室内,水泥墙慢慢地展现在眼前,沿着水泥墙走,最后可以看到海景。通道空间的流动性在二层空间处呈旋涡状向外伸出去,转移了人的视线,同时也让人关注这里的异彩纷呈:厨房角落设有一个小的潮汐式的水槽,卧室里有覆盖苔藓的岩石,从后面的庭院还能看到一片落叶树林。

钢框架屋顶和水泥墙及楼面的排列顺序一样。窄窄的自然光照射下来,投射出倾斜的光束。站在悬崖边,人们不会记得水泥地面有多牢固。钢框架木甲板悬浮在空中。透过无遮无拦的玻璃窗,远处海峡的美景尽收眼底;透过客厅地面上釉质的孔可以直接体验下面沙滩和大海带给人们的头晕目眩的感觉。建造过程中打磨的痕迹就像海滩上的漂浮物一般随处可见。

Tula House

Perched 13.4m above the Pacific Ocean on a remote island, the Tula House reflects the casual irregularity of the sites rock ledges, beach, and forest in both its geometric and spatial order.

The topography of the site is highly irregular; the prospects diverse. Views to the east stretch over the open water and islands of the Strait of Georgia to the mountain ranges on the mainland of British Columbia. Views to the south overlook a small tidal basin. Moss covered basalt hills are interspersed among treed expanses and richly vegetated crevices, valleys and swales. Stands of red alder and big-leafed maple enliven the predominantly dark Douglas fir forest. The shoreline below the house is littered with the flotsam and jetsam of the ocean where logs and rocks have been tossed around by the tides and storms like a child's game of "pickup sticks". One site is actually many sites.

The house cultivates a sense of dwelling with, and within, such diversity. Low rock walls edge a gravel approach to the house. A loose arrangement of concrete walls, clad in staggered fibre-cement panels, begins to describe space. These panels are black in color. From a distance, the house visually recedes into the dark forest. The roof, planted in moss and native ground covers, appears from above to be continuous with the surrounding ground plane. Groundwater flows continuously through the site where it is captured momentarily within an entry courtyard. The ground plane of the courtyard and interior floor of the house are large shard-like concrete plates. Within the house, spaces are defined by a series of slowly unfolding concrete walls that channel the flow of space through to ocean views. This primary flow is diverted in passage by eddies of secondary space which branch off, separating and focusing moments of diversity in the site: the small tidal basin off the kitchen nook, a ledge of moss covered rock in the bedrooms, a view back from the court to a swath of deciduous trees.

A steel-framed roof mirrors the order of concrete walls and floor plates. Narrow skylights project lines of light at oblique angles through space. At the cliff-edge, the solidity of concrete floors is left behind. A steel-framed, wooden deck is cantilevered into the air. While an uninterrupted expanse of glass takes in distant, sublime views of the straight, glazed apertures in the floor of the living room deck open vertiginously to the textures and sensual immediacy of beach and ocean below. Millwork elements float freely within the spaces like the flotsam and jetsam on the beach.

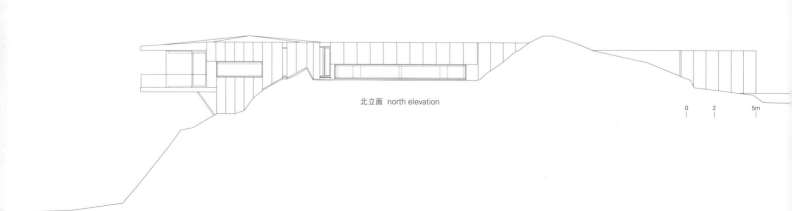

北立面 north elevation

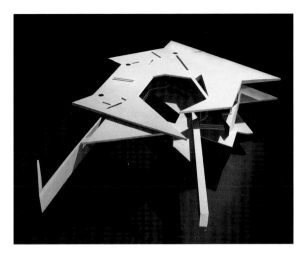
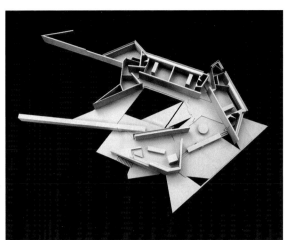

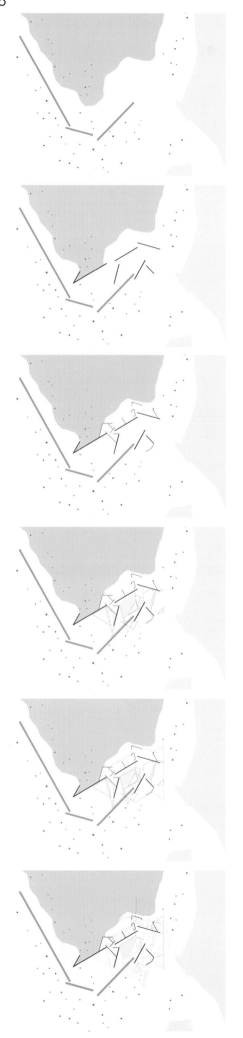

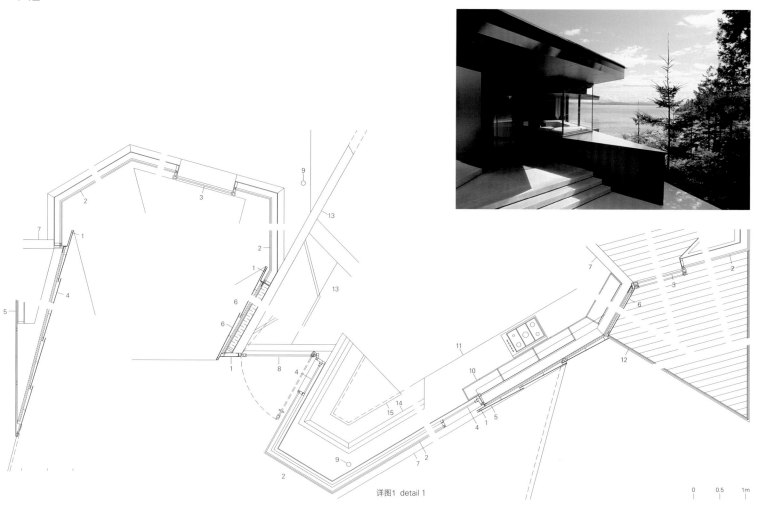

详图1 detail 1

1. painted metal plate assembly
2. aluminum curtain wall framed double insulated glass unit
3. sliding glass door
   thermally broken steel frame
   aluminum cladding
   overhead track system
4. aluminum framed window operator
   double insulated glass unit
5. cement board panel
   vertical angled metal subframing
   horizontal metal z-girts
   foil faced polyisocyanurate insulation
   self-adhered air vapour barrier
   plywood sheathing
   wood framing
   architectural concrete wall
6. cement board panel
   vertical angled metal subframing
   horizontal metal z-girts
   foil faced polyisocyanurate insulation
   self-adhered air vapour barrier
   architectural concrete wall
7. cast in place architectural concrete
8. hold open center hung door
9. structural steel column
10. architectural woodwork
11. cast in place concrete countertop
12. single layer tempered glass guard
13. interior garden

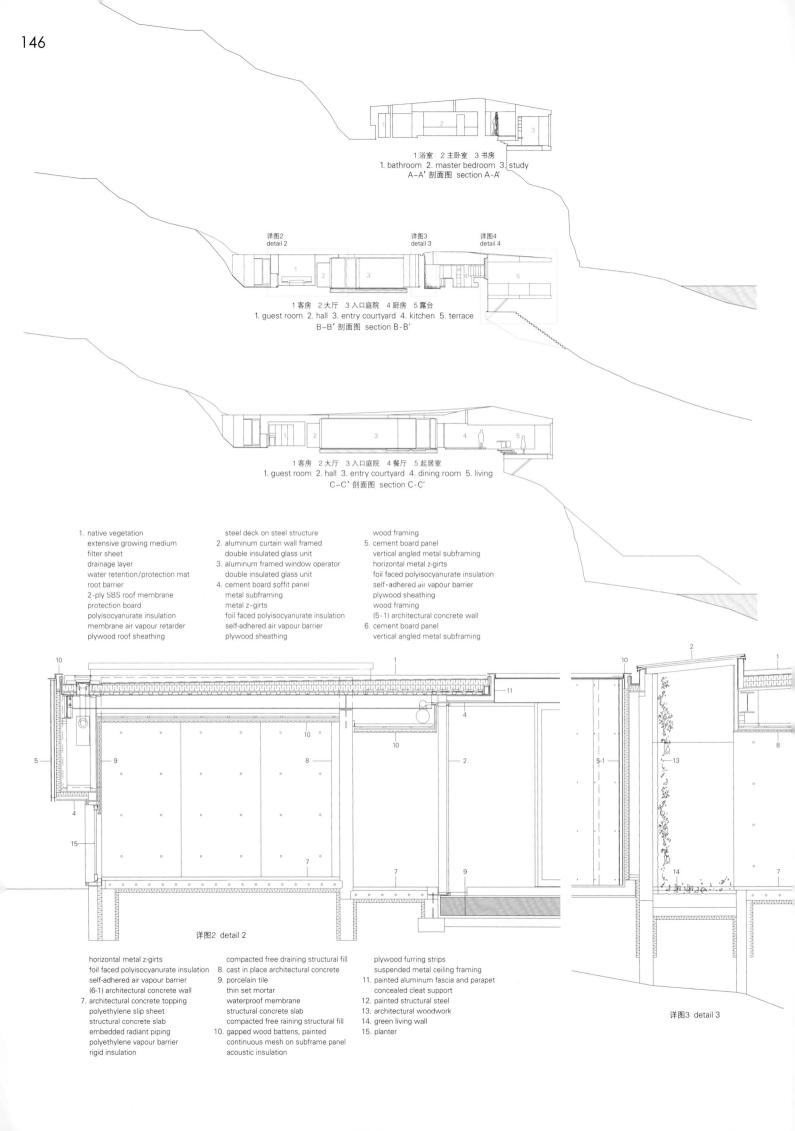

1 露台　2 书房　3 起居室
1. terrace　2. study　3. living
D-D' 剖面图　section D-D'

1 露台　2 起居室　3 书房
1. terrace　2. living　3. study
E-E' 剖面图　section E-E'

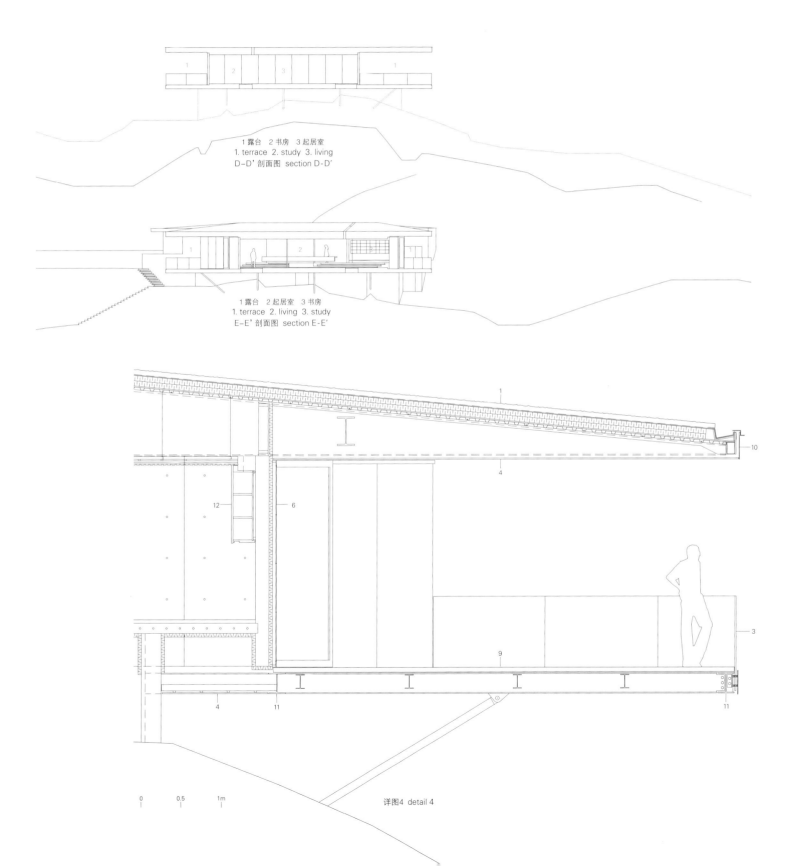

详图4　detail 4

独立式住宅

Pezo von Ellrichshausen Architects

这座独特的建筑实体位于乡间景观的显著位置内，那里种有葡萄和橄榄树，背对着中世纪的乡村和遍地露出地面的岩石。一处裸露在外的水平体量与地面分离开来，悬在古时代的背景中。

建筑是透明的，且浑然一体，位于一个看不见的基座上，极具平衡感，建筑的轮廓被一分为二，抬高的部分从远处可以看到，另一部分则被当地植物的叶子遮挡住了。平台上空展示了建筑最重要的部分。能够展现全貌的环形边界被16根柱子凸显出来，柱子按一定的间距排列，其间容纳的一系列房间的功能未定。这些水晶般的、对称的居住空间相互独立，通过拐角处的四个开放式露台连接起来，柱廊太窄，以至于无法

容纳一间静态的起居室。同时柱廊又太深，也无法建成一个观看平台的阳台。在平台上空，仅有的封闭空间（位于中心）没有屋顶；它的四面墙体在中心位置设有孔洞，地面是水（可能是人类能想到的最柔和的小径了），且反射了天空的场景。沿着一条笔直的倾斜的小路，台阶一分为二，形成了入口；进到房间的感觉像是走进了双子隧道，隧道环绕着中间的游泳池，只有小型洞口呈对角线状，允许人们的视线越过水面，望向天空。在入口水平面的下方，一条通道围绕着一间没有使用的房子（头顶有疏落的光线）。这个建筑体量的整体性让它自成一体；其具有纪念性的结构融在整体的轮廓里。正如智利理论家胡安·博彻斯所言，"建筑是物理的化身"，此话不仅指的是建筑的承重和张力，也是谈论生活本身，只是生活抹杀了人为的努力。

This unique entity occupies a dominant position in a rural landscape of vineyards and olive groves, set against a general background of medieval villages and rocky outcrops. A bare horizontal volume is detached from the ground, suspended in an almost archaic time.

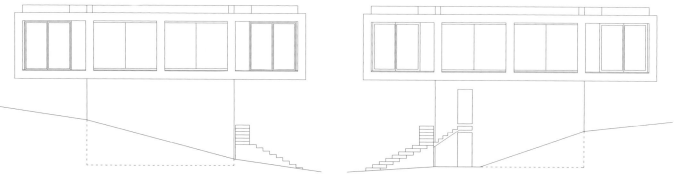

西南立面 south-west elevation 东北立面 north-east elevation

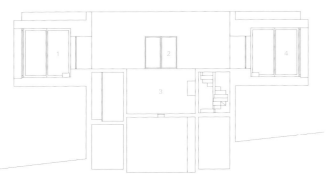

1 起居室 2 天井 3 泳池 4 卧室 1. living room 2. patio 3. pool 4. bedroom
A-A' 剖面图 section A-A'

1 露台 2 起居室 1. terrace 2. living room
B-B' 剖面图 section B-B'

Transparent and monolithic, the building is balanced on top of a blind pedestal, in such a way that its outline is divided between an elevated portion visible from a distance and another that disappears behind the leaves of native plant species. The aerial realm of the platform indicates the cardinal points. A panoramic perimeter ring punctuated by sixteen columns placed at regular intervals accommodates a sequence of rooms with undefined functions. These crystalline, symmetrical living spaces are mutually independent and are linked by four open terraces at their corners. This portico is too narrow to contain a static living room and too deep for a balcony – cum- viewing platform. In this aerial realm, the only closed room (in the center) is not roofed; its four walls are perforated at their central point, while the floor is of water (the gentlest paving known to man), which reflects the sky. After going along a sloping straight path, a bifurcated set of steps shapes the entrance; entering the house is like going into a twin tunnel that encircles the central swimming pool, with small openings set diagonally that allow glimpses of sky across the water. Under this access level a passageway goes around a room without use (and with scant overhead lighting). The volume unity has something of the generic schema about it; its monumental structure disappears in the outlines of its mass. It would seem that when Chilean theoretician Juan Borchers said that "architecture was physics incarnate," he was speaking not only of loads and tensions, but of life itself, which erases their effort.

项目名称：Solo House
地点：Cretas, Teruel, Spain
建筑师：Mauricio Pezo, Sofia von Ellrichshausen
合作建筑师：Diogo Porto, Bernhard Maurer, Valeria Farfan, Eleonora Bassi, Ana Freeze
建造商：Ferras Prats
甲方：Christian Bourdais, Solo House
用地面积：3,000m²
有效楼层面积：313m²
设计时间：2009~2010
施工时间：2010~2013
摄影师：©Cristobal Palma (courtesy of the architect)

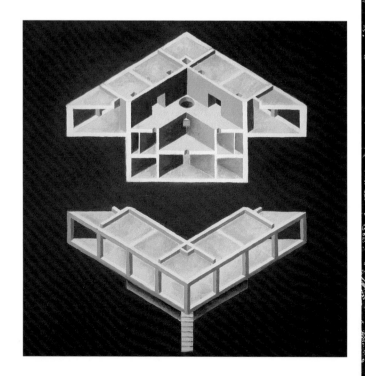

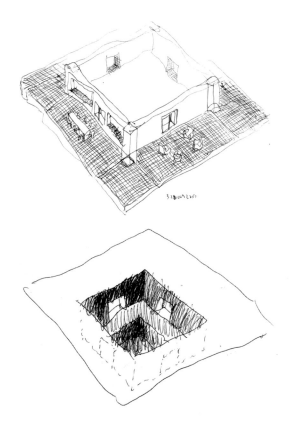

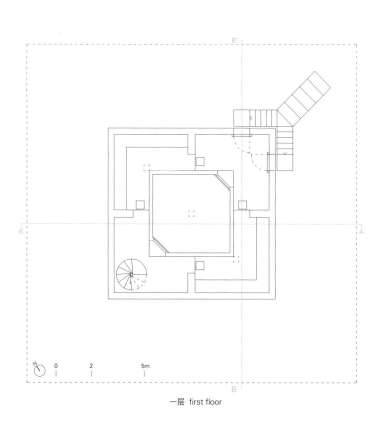

一层 first floor

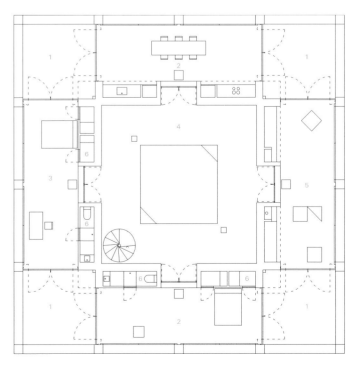

1 露台 2 餐厅 3 卧室 4 天井 5 起居室 6 浴室
1. terrace 2. dining room 3. bedroom 4. patio 5. living room 6. bathroom
二层 second floor

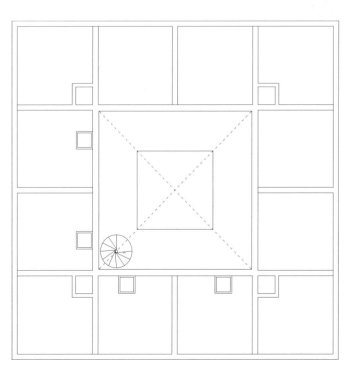

三层 third floor

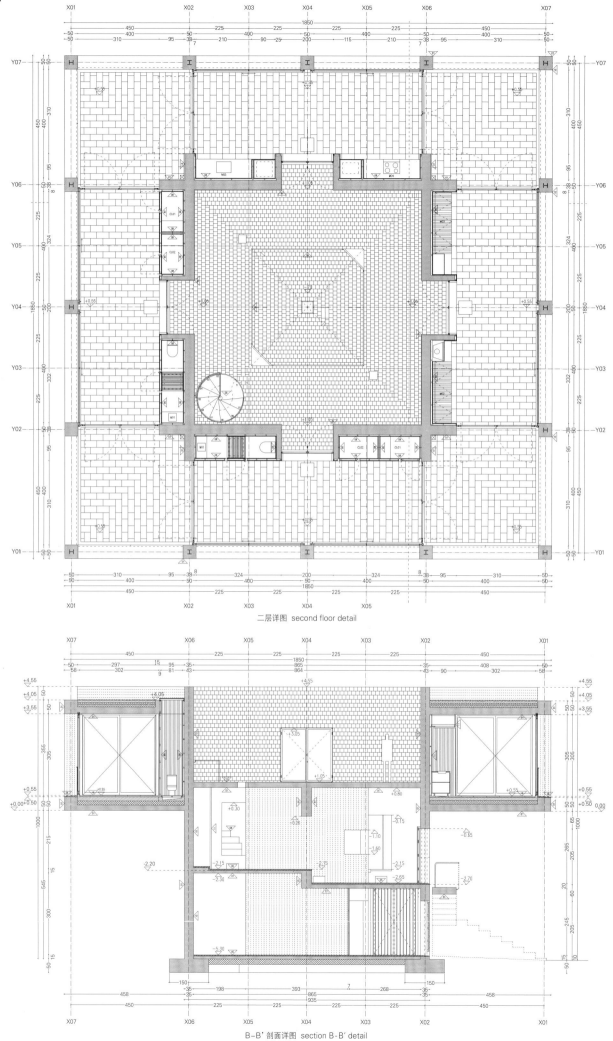

二层详图 second floor detail

B-B' 剖面详图 section B-B' detail

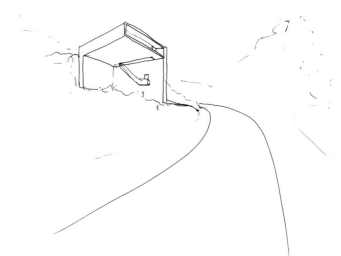

悬崖上的住宅
Fran Silvestre Arquitectos

我们喜欢建筑可以建在空中和水上的优点。一块土地俯瞰大海,如果真这样,最好的办法是什么都不做。即使这样,它也会让人流连忘返。可以考虑在这里建造一座尊重土地自然走势的建筑。从上往下,住宅和它的影子静静地俯瞰地中海。太阳底下,游泳池带我们走进海洋,这里成了安静的小海湾。变化的地方是楼梯,这里能唤起人们的回忆,另外,地下室还有一个花园。

由于地形陡峭以及住宅要建在同一水平上,所以根据此地的地貌特点,利用钢筋混凝土板和屏障搭建一个立体三维结构的方案被选中,如此一来也减少了土方工程。整块石头支撑的结构形成了一个从入口高度处开始的水平平台,住宅就建在上面。游泳池在较低的地方,以前这里也是平坦的。这个混凝土结构与外界隔离,外面粉饰着一层看起来柔软光滑的白水泥。出于对这里传统建筑的尊重,其他的材料、墙、人行道和屋顶的沙粒,这一切颜色都是一致的,强调颜色一致的同时,也突出了住宅的整体一致性。

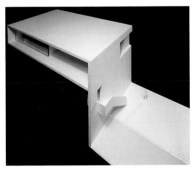
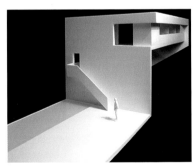
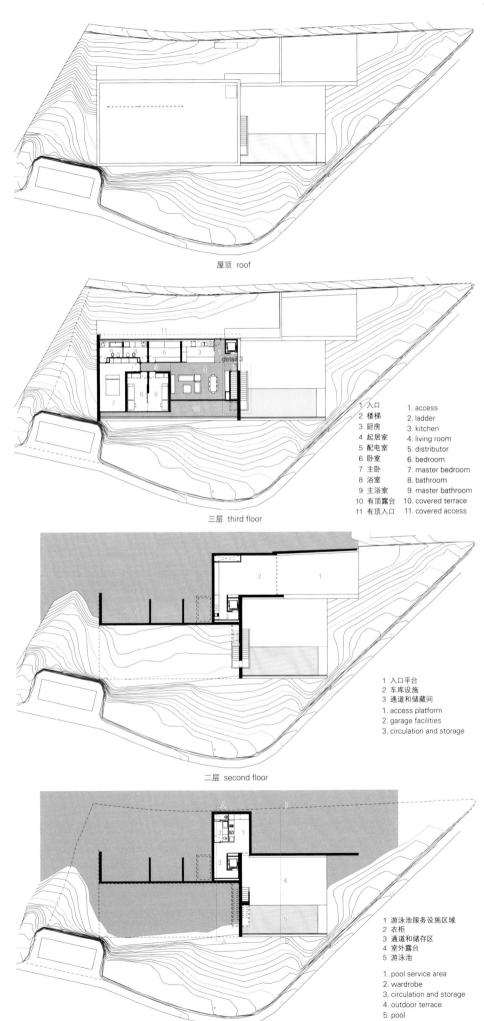

项目名称：House on Cliffside
地点：Toix Mascarat, Calpe, Alicante, Spain
建筑师：Fran Silvestre Arquitectos
项目团队：负责人_Fran Silvestre, María José Sáez/合作建筑师_Maria Masià, Adrián Mora, Jordi Martínez, José V. Miguel
合作者：Fran Ayala, Ángel Fito
结构工程师：David Gallardo
建筑工程师：Vicente Ramos, Esperanza Corrales, Javier Delgado
室内设计：Alfaro Hofmann
艺术指导：Adrián Mora, Maria Masià
承包商：Construcciones Alabort
用地面积：962.84m²
建筑面积：242m²
造价：EUR 650,000
竣工时间：2012
摄影师：©Diego Opazo (courtesy of the architect)

屋顶 roof

1 入口 — 1. access
2 楼梯 — 2. ladder
3 厨房 — 3. kitchen
4 起居室 — 4. living room
5 配电室 — 5. distributor
6 卧室 — 6. bedroom
7 主卧 — 7. master bedroom
8 浴室 — 8. bathroom
9 主浴室 — 9. master bathroom
10 有顶露台 — 10. covered terrace
11 有顶入口 — 11. covered access

三层 third floor

1 入口平台 — 1. access platform
2 车库设施 — 2. garage facilities
3 通道和储藏间 — 3. circulation and storage

二层 second floor

1 游泳池服务设施区域 — 1. pool service area
2 衣柜 — 2. wardrobe
3 通道和储存区 — 3. circulation and storage
4 室外露台 — 4. outdoor terrace
5 游泳池 — 5. pool

一层 first floor

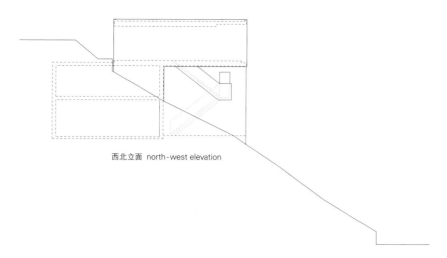

西北立面 north-west elevation

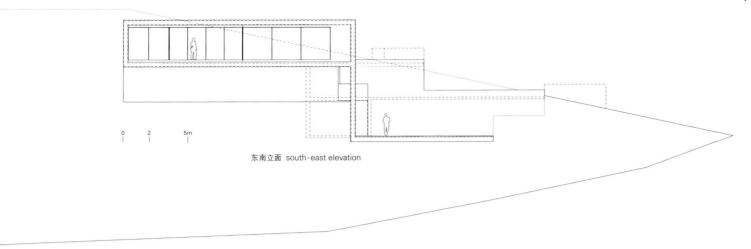

东南立面 south-east elevation

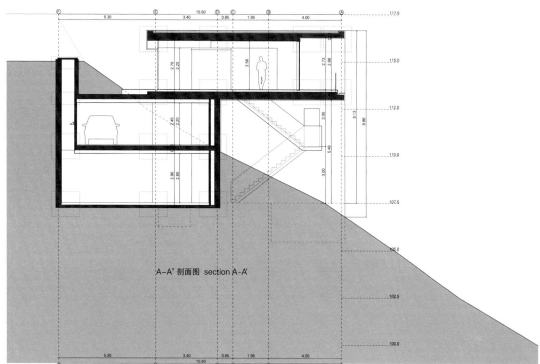

A-A' 剖面图 section A-A'

1. reinforced coping stone (slope 1%)
2. waterproofing membrane
3. concrete slab
4. bonding mortar + slate (external thermal insulation)
5. expanded polystyrene 4cm, type 3
6. anchors
7. double mesh between materials
8. outdoor board knauf
9. galvanized shaped steel
10. big raindrop
11. glass skylight (4 + 4; 6; 3 + 3)
12. wooden door
13. sheet metal stainless laced blanco
14. regularisation
15. non woven membrane + extruded polystyrene
16. poor mortar
17. non woven membrane + waterproofing membrane
18. vapour barrier
19. stone flooring + bonding mortar
20. natural stone
21. radiant floor
22. cellular concrete screed forming slope
23. gypsum plasterboard
24. garnished and coated with plasters

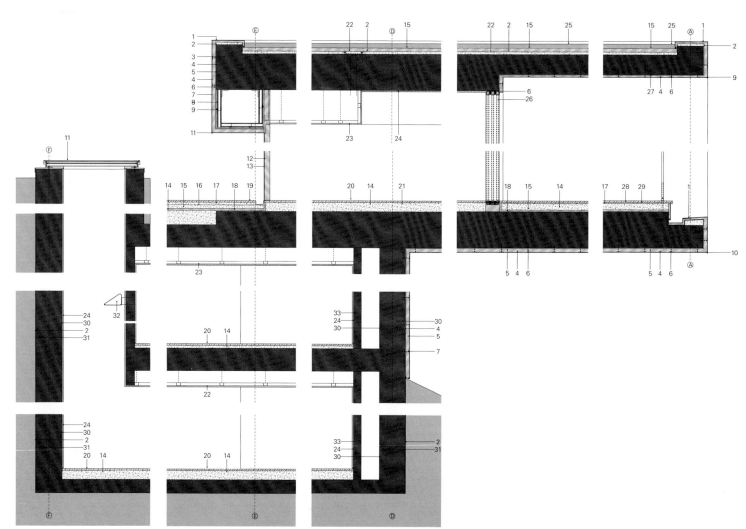

详图1_A-A' 剖面图 detail 1_section A-A'

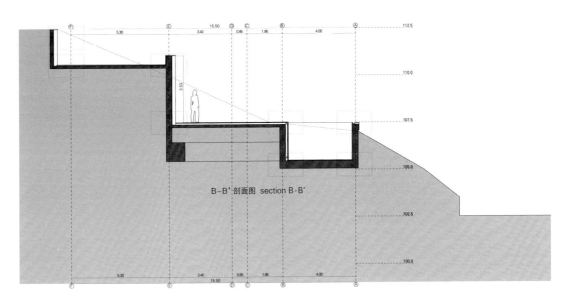

B-B' 剖面图 section B-B'

25. gravel
26. aluminum frames with break thermal profile
27. expanded polystyrene 2cm
28. flexible mortar (clemex)
29. stone floor
30. concrete wall 30cm
31. primer
32. accessible lamp

33. hollow brick 9cm
34. stone wall without metal anchors
35. catheter tubing
36. selected land
37. reinforced concrete slab
38. extruded polystyrene
39. non-woven membrane and drainage
40. land

41. coping stone
42. mortar grip polystyrene
43. projected concrete
44. mortar
45. reinforced and pigmented mortar
46. solid brick
47. hollow brick
48. chamber

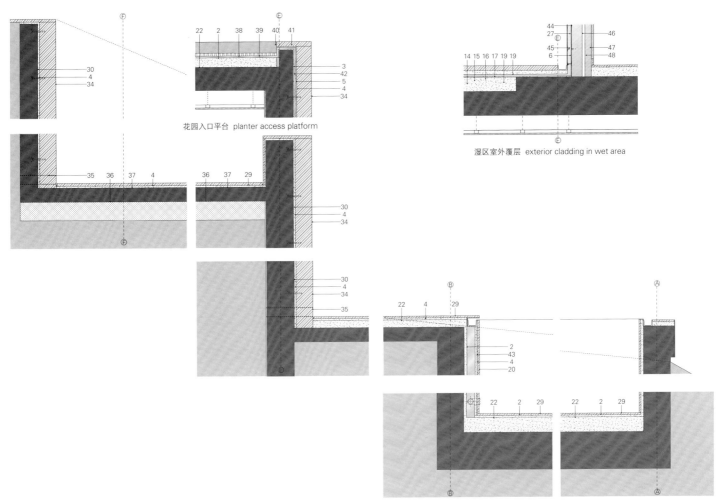

花园入口平台 planter access platform

湿区室外覆层 exterior cladding in wet area

详图2_B-B'剖面图 detail 2_section B-B'

House on the Cliff

We like the virtue of architecture which makes possible constructing a house on air, walking on water. An abrupt plot of land overlooking the sea, what is best is to do nothing. It invites to stay. A piece that respects the land's natural contour is set in it. Above, a shadow, the house itself, looks calmly at the Mediterranean. Under the sun, the swimming-pool brings us closer to the sea, it becomes a quiet cove. In the inflection point, the stairway proposes an evocative path, a garden in the basement.

Due to the steepness of the plot and the desire to contain the house in just one level, a three-dimensional structure of reinforced concrete slabs and screens adapting to the plot's topography was chosen, thus minimizing the earthwork. This monolithic, stone-anchored structure generates a horizontal platform from the accessing level, where the house itself is located. The swimming-pool is placed on a lower level, on an already flat area of the site. The concrete structure is insulated from the outside and then covered by a flexible and smooth white lime stucco. The rest of materials, walls, pavements, the gravel on the roof...all maintain the same colour, respecting the traditional architecture of the area, emphasizing it and simultaneously underlining the unity of the house. Fran Silvestre Arquitectos

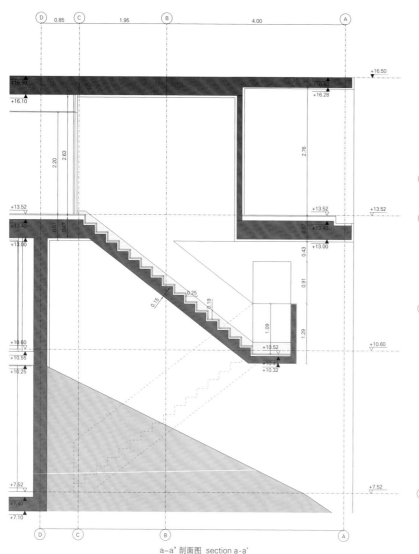

a-a' 剖面图 section a-a'

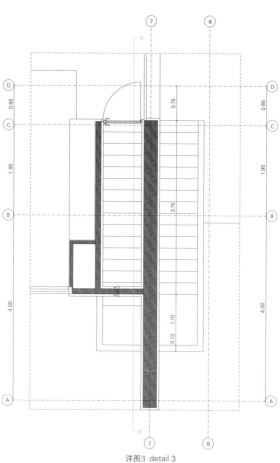

详图3 detail 3

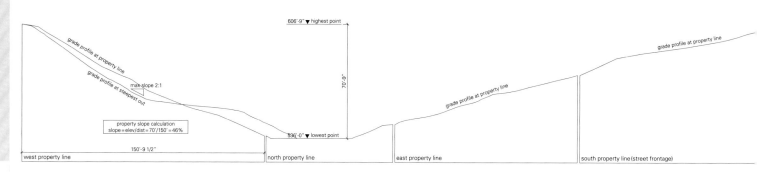

车库房
Anonymous Architects

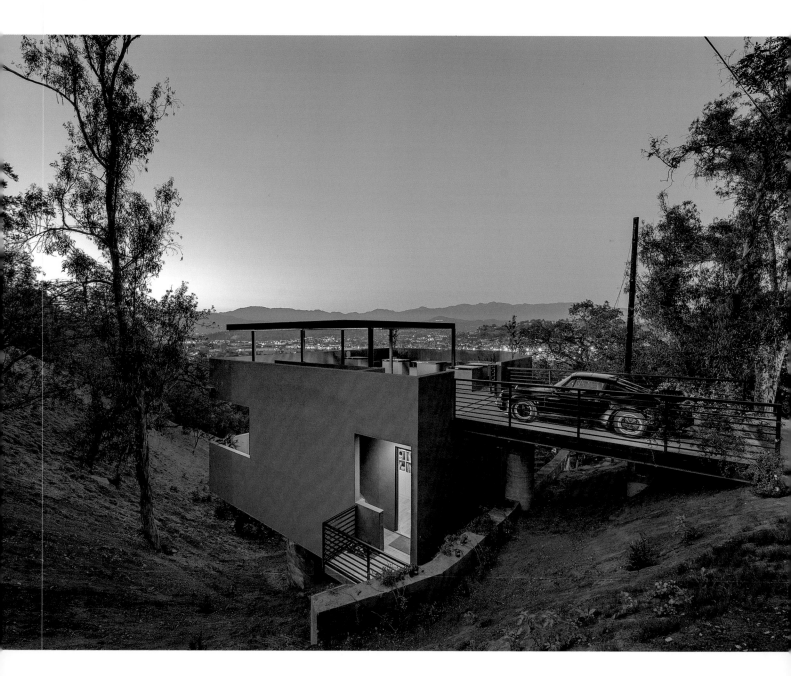

临街空地，顺势朝下，极其陡峭，房子的车库建在屋顶，而人住楼下。除了车上人下颠覆传统的设计别出心裁外，设计还能满足停放两辆车的要求。

倒置的设计使房子一楼的功能搬到了楼顶，简单的开车回家这一行为因为每次都要开到屋顶，因而每次都会让人感到新奇。

同时，屋顶可以像甲板一样利用，站在上面可以一览位于洛杉矶东北方向的圣加布里埃尔山的景致。

因为地形陡峭，房子像是飘浮在山中。既减少了打造基础的工作量，同时也意味着连接桥是进入房间的唯一通道，如此一来就真正成为悬浮建筑了。

停车处和房子建在一起的另一大好处是省去了另建车库再打基础和修墙的麻烦。

Car Park House

Starting with a vacant lot with a very steep down-slope from the street, the design of the house places the carport on the roof with the residence below. In addition to being a dramatic shift of expectations, it is also a logical response to the building code which requires parking for two vehicles.

This inversion moves the typical ground floor of the house up on the roof and makes the simple act of arriving home – and driving onto the roof of the house – a surprise every time.

The roof is also usable as deck space and has unobstructed views of the San Gabriel Mountains, which are to the Northeast of Los Angeles.

Because of the steep terrain the house is designed to float over the hillside. This reduces the amount of foundation required and also means that the only way to access the house is over the bridge – so it is truly a floating structure.

The added benefit of providing the parking and the house as the same structure is to eliminate the need for additional foundations and walls for a garage.

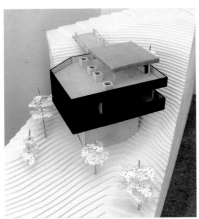

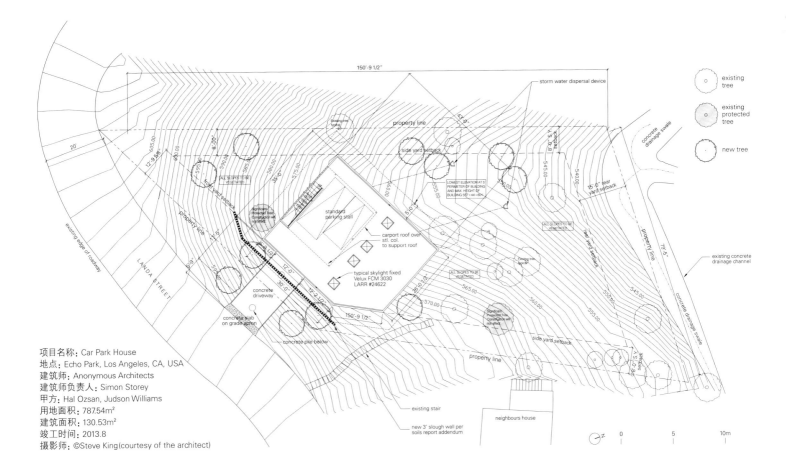

项目名称：Car Park House
地点：Echo Park, Los Angeles, CA, USA
建筑师：Anonymous Architects
建筑师负责人：Simon Storey
甲方：Hal Ozsan, Judson Williams
用地面积：787.54m²
建筑面积：130.53m²
竣工时间：2013.8
摄影师：©Steve King (courtesy of the architect)

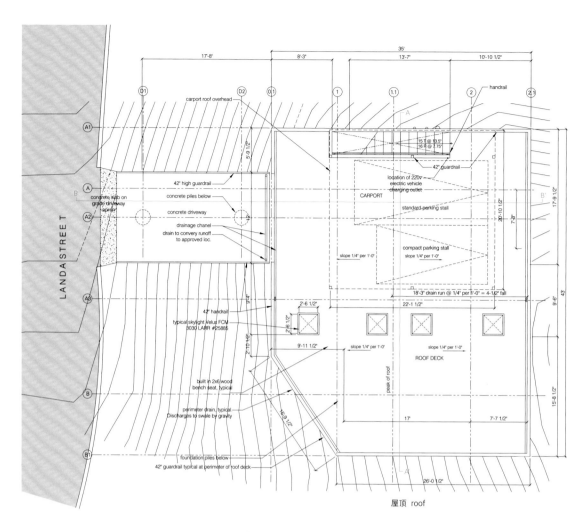

屋顶 roof

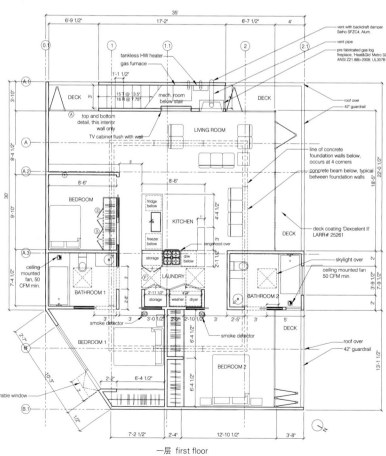

一层 first floor

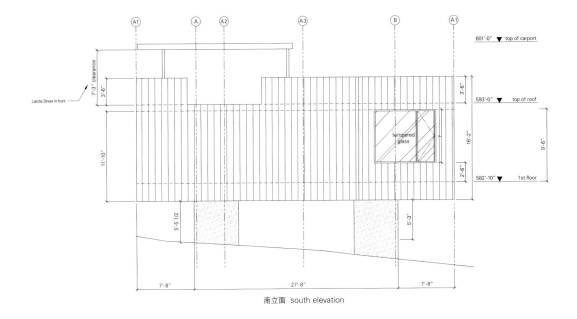

南立面 south elevation

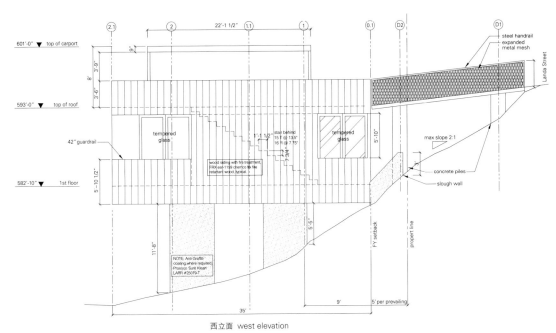

西立面 west elevation

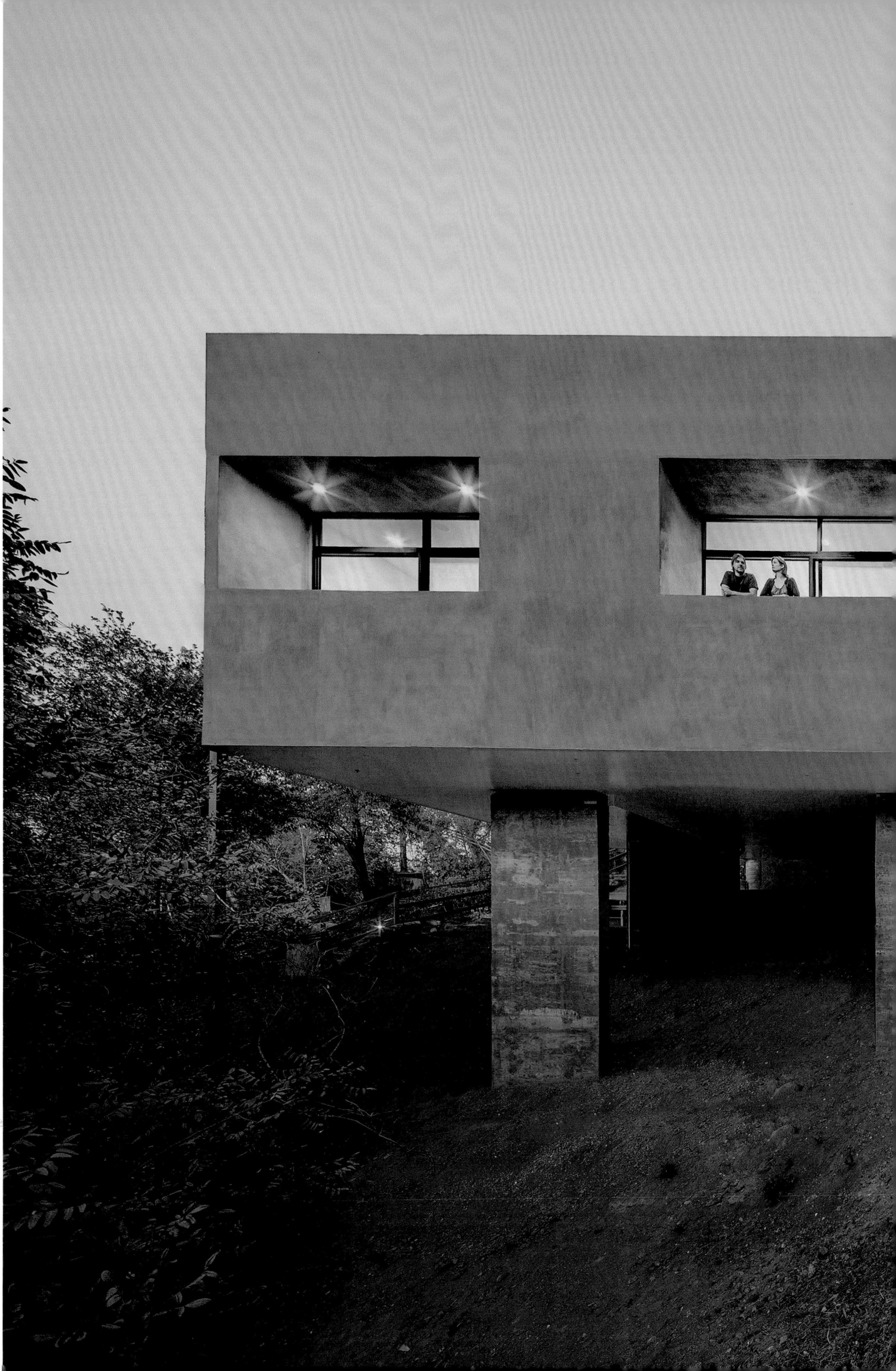

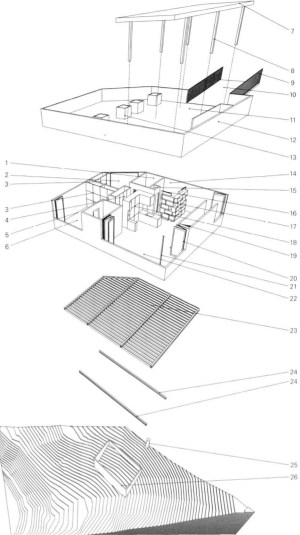

1. 2nd bathroom, accessible from bedroom and hall
2. 2nd bedroom
3. closet panelized doors, painted MDF
4. hallway closet, bifolding doors: standard
5. master bathroom
6. master bedroom, 5/8" hardwood flooring
7. a. expanded metal, b. sheet metal
8. steel pole, painted
9. custom steel handrail
10. driveway: a. 3" concrete / b. 2 piles 16" diameter
    note: not connected to house
11. a. lightweight concrete / b. steel decking / c. 2"x14" joists /
    d. waterproofing above and below concrete
12. solid handrail, interior: decking, exterior: wood siding
13. skylight masked as planter
14. office/guest bedroom
15. office closet
16. office pocket doors, full height 9' 6"
17. wall mounted shelving unit with sliding doors: 3/4" hardwood
18. kitchen cabinets, painted MDF
19. entrance/staircase 5'/8" hardwood
20. bifolding exterior doors, Cantina
21. steel post
22. interior: 5/8" hardwood flooring,
    exterior: 3/4" decking,
    match heights
23. 1.5"/12" glue lam, floor joists
24. 6"/12" I beam, steel
25. 16" concrete pile
26. concrete pier, on stepped foundation

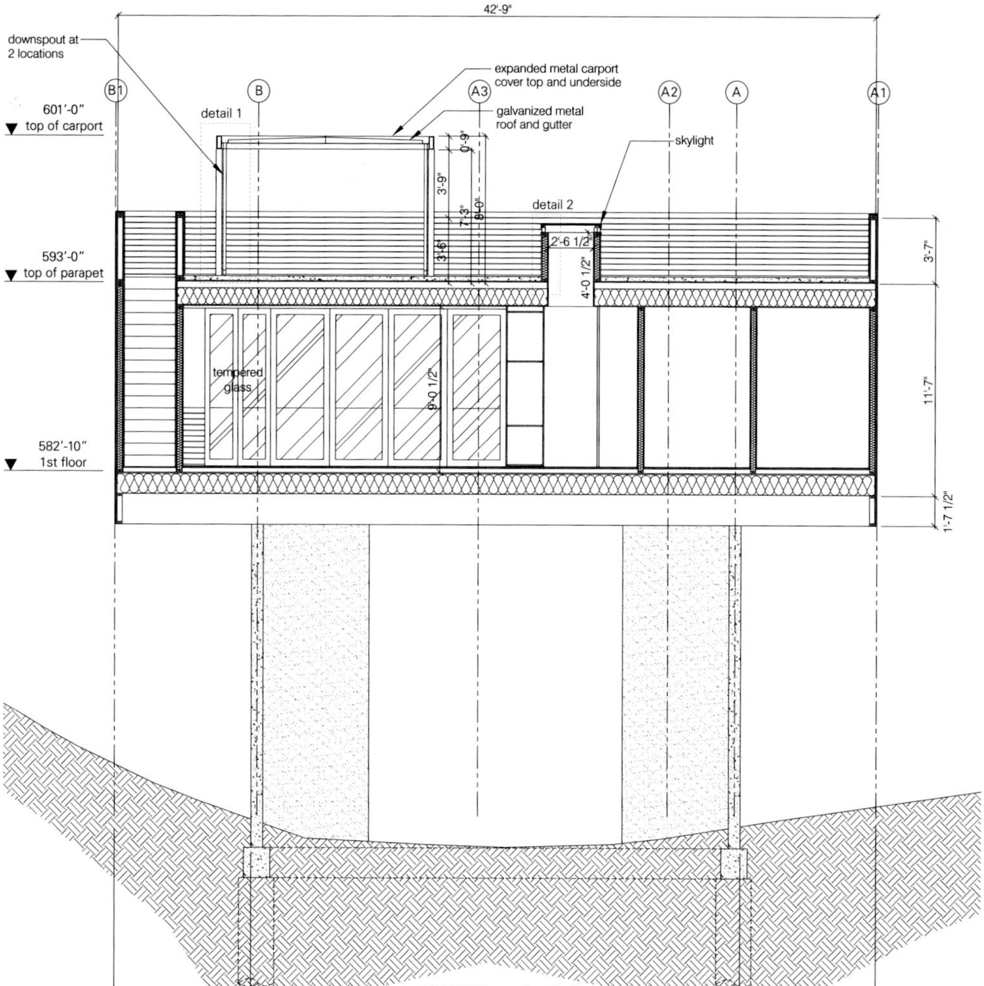

A-A' 剖面图 section A-A'

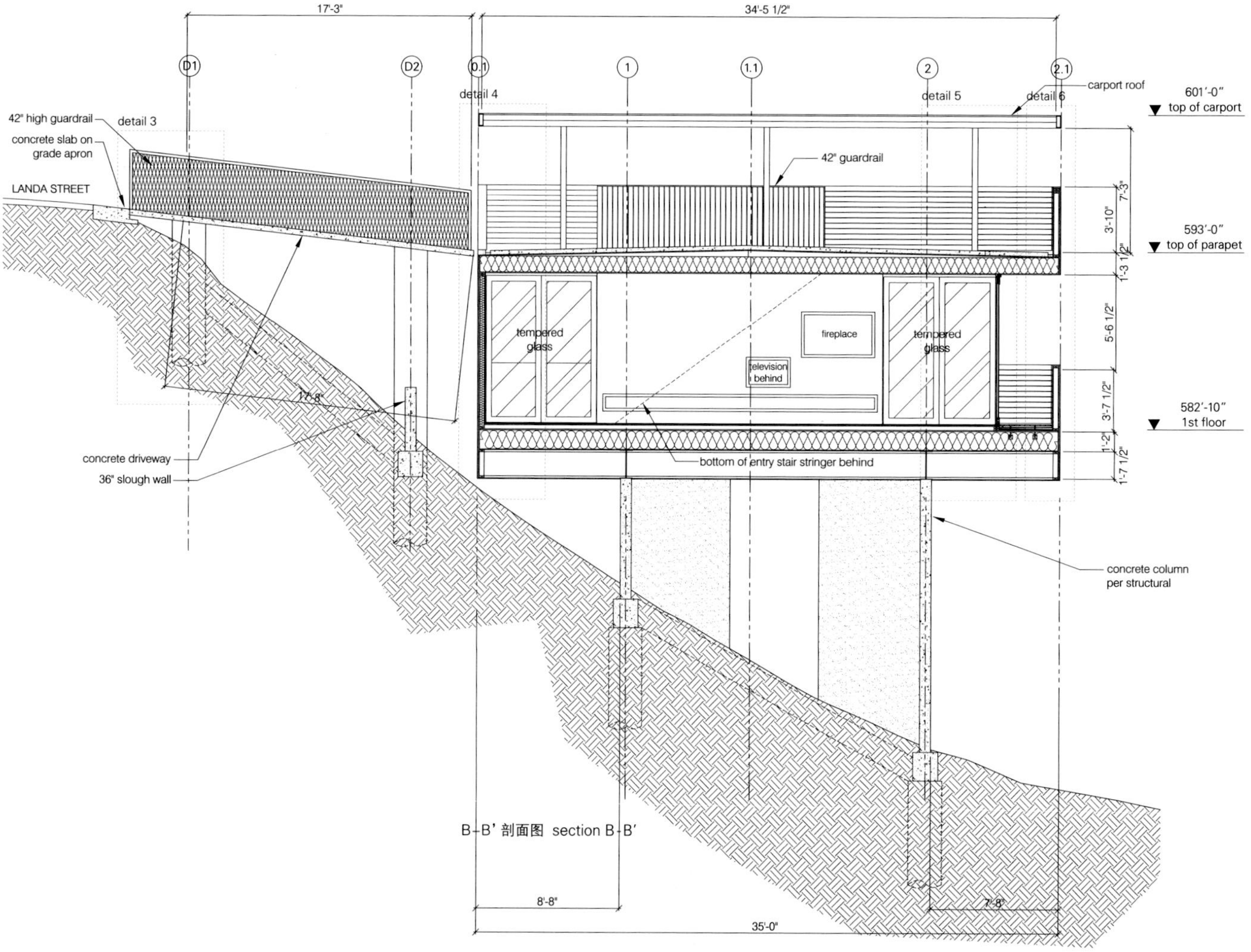

B-B' 剖面图 section B-B'

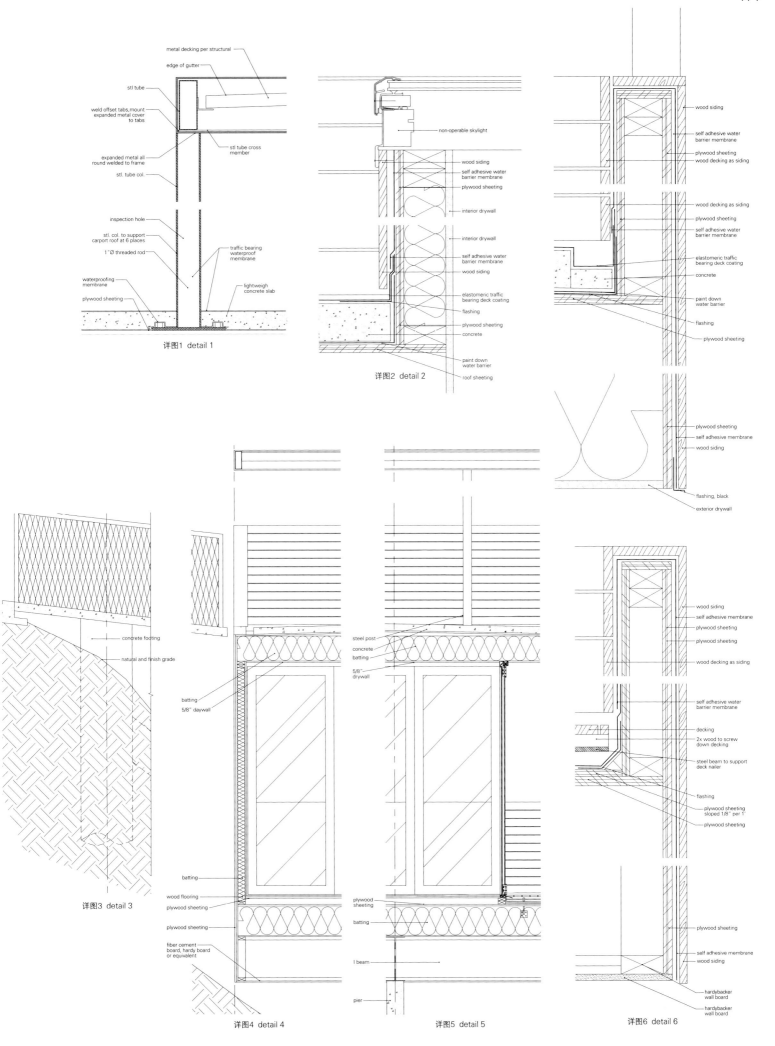

住宅 艺术高度 Dwell How Artificially High

石屋
Olson Kundig Architects

主人钟情于建造房屋时保留裸露在外的石头，这激发了建筑师的设计灵感。Pierre即古法语中的岩石，建筑师在进行设计构思时想把它建成一个舒适地偎依在岩石中的避风港，石屋的建成凸显了这个地方的重要性。"在岩石上的房子选址是在这片土地中最贫瘠的地方，其他地方留作耕地用，"昆丁是这样说的。因为用料粗制，房屋由石头围建而成，外加绿屋顶和周边的植物，所以从特定的角度来看，房屋消失在了自然之中。

为了让房屋基础牢固，突兀的岩石被机器和人工挖掉了。电流接触器配上大型钻头勾勒出建筑的轮廓，然后用炸药、液压切片机、钢丝锯和其他工具作业，施工越往后，对工具的要求也就越精细。被挖出的岩石粉碎后被当做集料来制造混凝土楼板。所有的砌石工程上面都留有挖凿的痕迹，标志着建造的过程，而停车场都是利用大块岩石建成的。

除了独立的客房外，房屋内开放式厨房、餐厅和起居室这些发挥重要功能的地方都集中在一层。木质的储物箱（利用从Lionel Pries设计的房屋中回收而来的侧板制造而成）完成了房屋从室外到室内的过渡。两个巨大的书橱打开的话，就成为洗衣房和厨房储食处的隐藏式入口。打开一扇大型的钢玻旋转门可以到达户外的露台。主人套间位于主要空间的一个直角位置，其中摆放一张特制的床，床头板以皮革制成，床尾板嵌入位于房屋中间的连接地板和天花板的书架中。

纵观整座房屋，墙上岩石突出，与豪华的家装风格形成鲜明的对比。火炉内外都是用现成的石头雕刻的，唯独火炉顶平整，其他地方都未被打磨过。主卫生间里水像瀑布一样从三个打磨光滑的池子流下来，主卫的洗涤槽也是用天然石头打造出来的。在室内不起眼的地方还有一个用岩石打造出来的化妆室，镜子被镶嵌在采光井上，正好把自然光反射到室内。

建筑结构的材质——低碳钢、光滑混凝土和干板墙——成为内部装饰、艺术品和从外部欣赏到的海湾和周围环境的天然背景。卡梅伦·马丁、耶西·保罗·米勒、安德烈·塞拉诺、弗朗兹·韦斯特和克劳德·泽瓦斯等人的当代艺术品挂在房屋内外。古董家具、艺术品和定制物品摆在一起相得益彰。照明设备是根据西雅图艺术家和电灯设计师艾琳·麦高文的设计定制的，艾琳·麦高文因和著名的建筑师罗兰·特里合作而被世人熟知。

The Pierre House

The owner's affection for a stone outcropping on her property inspired the design of this house. Conceived as a retreat nestled into the rock, the Pierre (the French word for stone) celebrates the materiality of the site. "Putting the house in the rock follows a tradition of building on the least productive part of a site, leaving the best parts free for cultivation," Kundig says. From certain angles, the house – with its rough materials, encompassing stone, green roof, and surrounding foliage – almost disappears into nature.

To set the house deep into the site, portions of the rock outcropping were excavated through a combination of machine work and handwork. The contactor used large drills to set the outline

of the building, then used dynamite, hydraulic chippers, and wire saws and other hand tools, working with finer and finer implements as construction progressed. Excavated rock was reused as crushed aggregate in the concrete flooring. Excavation marks were left exposed on all the stonework, a reminder of the building process, while huge pieces of rock were employed for the carport structure.

With the exception of a separate guest suite, the house functions on one main level, with an open-plan kitchen, dining, and living space. A wood-clad storage box (made with siding reclaimed from a Lionel Pries–designed house) transits from outside to inside. Its two large bookcases open to provide concealed access to laundry and kitchen storage. A large pivoting steel and glass door provides access to a terrace. Set at a right angle to the main space, a master suite features a custom-designed bed with a leather headboard and footboard set in the middle of floor-to-ceiling bookshelves. Throughout the house, the rock protrudes into the space, contrasting with the luxurious textures of the furnishings. Interior and exterior fireplace hearths are carved out of existing stone; leveled on top, they are otherwise left raw. In the master bathroom, water cascades through three polished pools, natural sinks in the existing stone. Off the main space, a powder room is carved out of the rock; a mirror set within a skytube reflects natural light into the space.

The materiality of the built structure – mild steel, smooth concrete, and drywall – creates a neutral backdrop for the interior furnishings and artwork and the exterior views to the bay and surrounding landscape. Contemporary works of art by Cameron Martin, Jesse Paul Miller, Andres Serrano, Franz West, and Claude Zervas are mounted inside and outside the house. Antique furniture and art objects are complemented by custom pieces. The custom light fixtures are based on the designs of Irene McGowan, a Seattle artist and lighting designer best known for her work with noted architect Roland Terry.

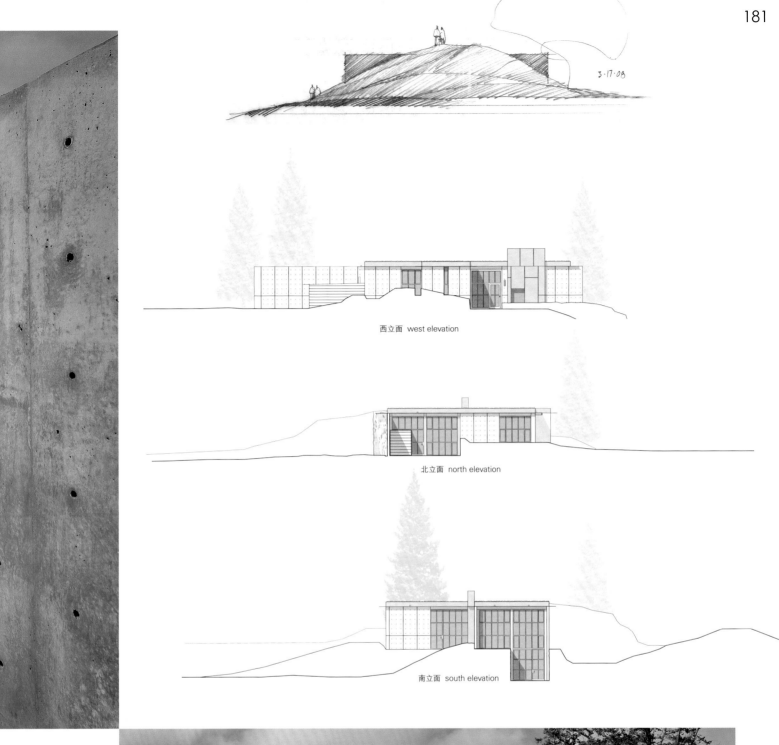

西立面 west elevation

北立面 north elevation

南立面 south elevation

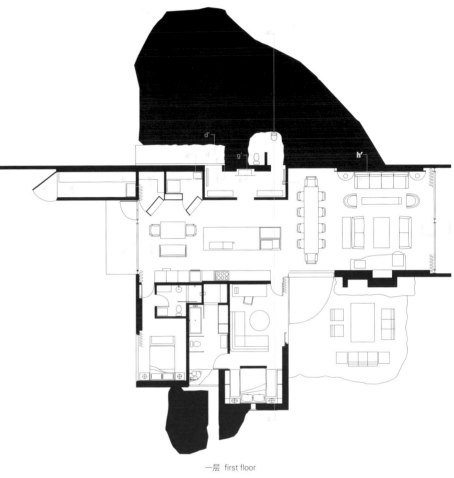

一层 first floor

项目名称：The Pierre
地点：Lopez Island, Washington
建筑师：Olson Kundig Architects
项目团队：设计负责人 _ Tom Kundig
项目经理 _ Chris Gerrick
室内设计 _ Charlie Fairchild
结构工程师：MCE Structural Consultants
土木工程师：CPL
地质技术工程师：Associated Earth Sciences
总承包商：Schuchart/Dow Construction
总建筑面积：232m²
施工时间：2008.12~2010.8
摄影师：©Benjamin Benschneider(courtesy of the architect) -
p.178, p.181, p.183, p.185, p.188, p.189
©Dwight Eschliman(courtesy of the architect) - p.180, p.182~183

A-A' 剖面图 section A-A'

a-a' 剖面图，入口处
section a-a', at door

b-b' 剖面图，储藏间处
section b-b', at storage box

c-c' 剖面图，起居室处
section c-c', at living room

d-d' 剖面图，入口门处
section d-d', at entry door

e-e' 剖面图，西侧起居室处
section e-e', at west living room

f-f' 剖面图，盥洗室处
section f-f', at powder room

g-g' 剖面图，浴室处
section g-g', at MW bath

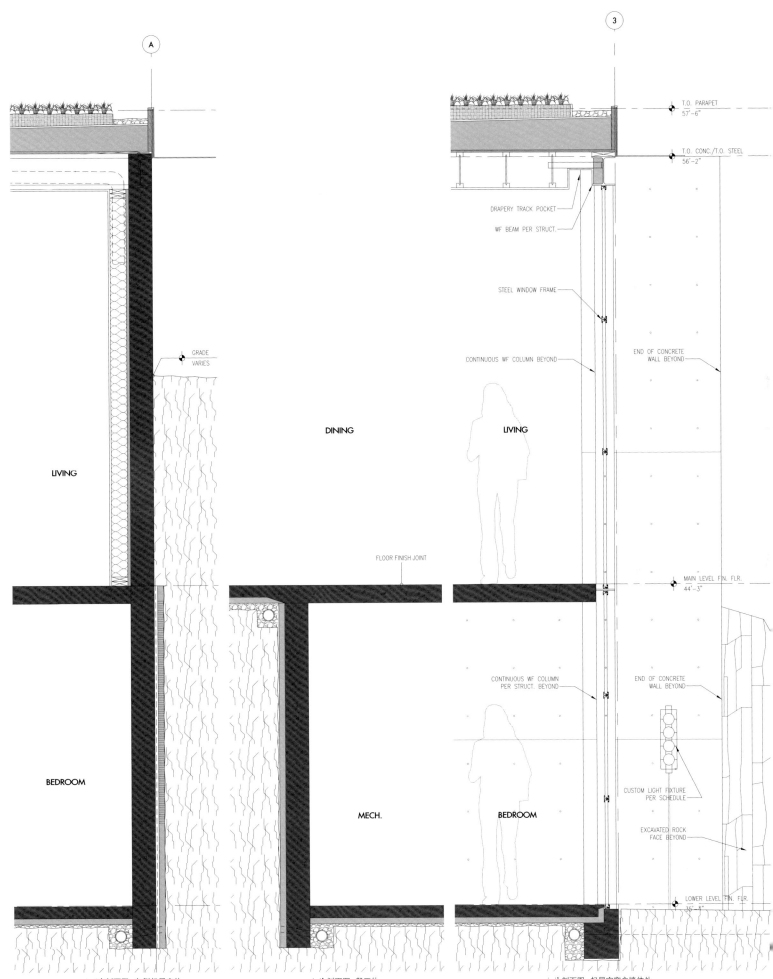

>>86
Zermani Associati Studio di Architettura
Paolo Zermani is a founder of Zermani Associati Studio di Architettura. Is also a professor of Architectural Composition at the Faculty of Architecture in Florence. Was an executive director of the international architecture magazine "Materia" from 1990 to 2000. Also has a significant production of projects published in several architectural magazines including A+U, Architectural Review etc. Has been invited to the Venice Architecture Biennale in 1991, 1992, 1996 and to the Milan Triennale in 1995 and 2003. In 2004, the Space Gallery in New York organized the exhibition "Paolo Zermani, Architecture in the Italian Landscape".

>>52
Dethier Architecture
Daniel Dethier studied architecture and civil engineering at the university and graduated in 1979. Founded Dethier Architecture in 1992. Free of all prejudice, their procedures are characterized by a critical vision allowing them to provide innovative solutions appropriate to current requirements in fields as varied as the type of habitat in towns and in the country, adding value to our heritage, cultural infrastructures, development of public areas, urban planning, design of interiors, furniture or equipment.

>>118
Caples Jefferson Architects
Sara Caples[left] is a principal of CJA. Graduated from Smith College with cum laude and received a M. Arch from Yale University. Has been working in CJA for 22 years. Everado Jefferson[right] is also a principal of CJA. Graduated from Pratt Institute with BID and Yale University with M. Arch. Has been working in CJA for 24 years.

>>36
Andrea Dragoni Architetto
Andrea Dragoni was born in Perugia, Italy in 1969 and obtained his degree at the Faculty of Architecture of Florence. Has been a contract professor at the Polytechnic University of Milan and also at the Faculty of Engineering of Perugia University. Is currently teaching at the Pietro Vannucci Fine Arts Academy in Perugia. Works in the fields of architecture, engineering and design. Has published articles in the most important architecture magazines including Bauwelt of Germany and has participated in important group exhibitions.

>>148
Pezo von Ellrichshausen Architects
Is an art and architecture studio established in Concepcion, southern Chile in 2002 by Mauricio Pezo[left] and Sofia von Ellrichshausen[right]. They have been the curators of the Chilean Pavilion at the 11th Venice Biennale in 2008. Also in 2010, they were invited by Kazuyo Sejima to the official selection at the 12th Venice Biennale. They teach regularly in Chile and have been visiting professors at the University of Texas, Austin and at Cornell University. Their works have been edited in monographic issues of A+U, 2G and ARQ.

>>62
Batlle i Roig Arquitectes
Both Enric Batlle i Durany and Joan Roig i Duran were born in Barcelona in 1956 and 1954, respectively. They studied at ETSAB and qualified as architects in 1981. In the same year, they set up Batlle i Roig Arquitectes and won the competition for the Roques Blanques cemetery in El Papiol, Barcelona which became a practice's first built work. Since then, they have produced many projects in the fields of architecture, landscape and urban planning.

>>112
Davis Brody Bond Architecture
Founded in 1952, the firm's work has combined social responsibility with design excellence.
Peter D. Cook received B. Arts from Harvard University and M. Arch from Columbia University. Founded Davis Brody Bond's Washington DC office in 2005. Recently he has resigned David Brody Bond. Robert Anderson who is a director of Davis Brody Bond's Washington DC office received a B. Arch from Washington State University. Has 18 years of experience working in the United States and internationally.

>>160
Fran Silvestre Arquitectos
Was founded in Valencia in 2005 by Fran Silvestre who was born in 1976. After having graduated in Architecture at the Superior Technical School of Architecture of Valencia in 2001, he studied urban planning at the Eindhoven University of Technology a year later. He has been teaching at the Polytechnic University of Valencia since 2006, and the European University since 2009. The firm's projects have been published in international architectural magazines and exhibited in some famous museums and galleries.

Heidi Saarinen
Was born in Finland, raised in Sweden. Has extensive teaching experience and project management, with specialism in experiential methodologies and spatial analysis in teaching and learning, encouraging students to interact with space through movement; questioning behavior and use of space. Is currently working on collaborative projects linking film, digital media, architecture and design, community and narrative space based in London. Is also a senior lecturer in Interior Architecture and Design at the University of Hertfordshire, and also eternal advisor at University of the Arts London and external examiner at Coventry University, UK.

Diego Terna
Received a degree in architecture from the Politecnico di Milano and has worked for Stefano Boeri and Italo Rota. Has been working as critic and collaborating with several international magazines and webzines as editor of architecture sections. In 2012, he founded an architectural office, Quinzii Terna together with his partner Chiara Quinzii. Currently is an assistant professor of Politecnico di Milano and also runs his personal blog L'architettura immaginata(diegoterna. wordpress.com).

Silvio Carta
Is an architect and researcher based in London. Received Ph.D. from University of Cagliari, Italy in 2010. His main fields of interest is architectural design and design theory. His studies have focused on the understanding of the contemporary architecture and the analysis of the design process. He taught at the University of Cagliari, Willem de Kooning Academy of Rotterdam and Delft University of Technology. He is now senior lecturer at the University of Hertfordshire. Since 2008 he is editor-at-large for C3 and his articles have also appeared in A10, Mark, Frame and so on.

>>74

Mancini Enterprises
Was founded in 2004. Offers comprehensive design services in the fields of urban planning, architecture, interiors, landscape and product design. With base in Chennai, South India, Mancini's team of 28 professionals headed by J.T. Arima[top] and Niels Schoenfelder[bottom] currently design urban master plans, hotels, resorts, schools, residential developments, residences, gardens, interiors, furniture and lighting for international clients.

>>100

Balonas & Menano
Pedro Balonas[left] was born in Coimbra, Portugal in 1966 and established his own firm in 1998. In 2007, he associated with Miguel Menano and developed some projects of high complexity, continuously focused on innovation and incorporation of new technologies. Simão Silva[right] was born in Oporto, Portugal in 1977 and graduated from the Faculty of Architecture - University of Porto in 2001. Joined Balonas & Menano in 2003 as a design architect. Has been a team leader and head of a department since 2008.

Nelson Mota
Graduated from the University of Coimbra in 1998 and received a master's degree in 2006 where he lectured from 2004 to 2009. Was awarded the Távora Prize in 2006 and wrote the book called A Arquitectura do Quotidiano in 2010. Is currently a researcher and guest lecturer at the TU Delft, in the Netherlands. Is also a member of the editorial board of the academic journal Footprint and also one of the partners of Comoco Architects.

>>170

Anonymous Architects
Simon Storey was born in New Zealand in 1971 and left for the United States in 1993. Graduated from Southern California Institute of Architecture(SCI-Arc) with a Masters of Architecture in 2004 and established his architectural practice in 2005 in Los Angeles, California. The practice focuses mostly on single and multi family residential and commercial in the Los Angeles area, although recently started projects located in NYC and North Carolina.

>>178

Olson Kundig Architects
Began its creative existence with Jim Olson. Tom Kundig, fellow of the American Institute of Architects, joined Olson as principal and owner in 1986. He received Bachelor of Arts in Environmental Design in 1977 and Master of Architecture from the University of Washington with Magna Cum Laude in 1981. The works of Olson Kundig have received over 50 awards from the American Institute of Architects and appeared in hundreds of publications worldwide.

>>134

Patkau Architects
Was founded in 1978 by John Patkau^{left} and Patricia Patkau^{right}. Is an innovative architecture and design research studio based in Vancouver, British Columbia, and Canada. Projects vary in scale from gallery installations to master planning, from modest houses to major urban libraries. As a design leader at Patkau Architects, John has instigated and developed the design of a wide variety of project types for a diverse range of clients nationally and internationally. Whereas Patricia has made important contributions to the field architecture in both practice and education. She is Emerita Professor at the University of British Columbia.

C3, Issue 2014.3
All Rights Reserved. Authorized translation from the Korean-English language edition published by C3 Publishing Co., Seoul.

© 2013大连理工大学出版社
著作权合同登记06-2014年第072号

版权所有·侵权必究

图书在版编目(CIP)数据

殡仪类建筑：在返璞和升华之间：汉英对照 / 韩国C3出版公社编；王平等译. —大连：大连理工大学出版社，2014.5

(C3建筑立场系列丛书；39)

书名原文: C3 Funeral between Nature and Artefact

ISBN 978-7-5611-9110-1

Ⅰ. ①殡… Ⅱ. ①韩… ②王… Ⅲ. ①丧葬建筑－建筑设计－汉、英 Ⅳ. ①TU251.6

中国版本图书馆CIP数据核字(2014)第088799号

出版发行：大连理工大学出版社
　　　　　（地址：大连市软件园路80号　邮编：116023）
印　　刷：上海锦良印刷厂
幅面尺寸：225mm×300mm
印　　张：12.25
出版时间：2014年5月第1版
印刷时间：2014年5月第1次印刷
出 版 人：金英伟
统　　筹：房　磊
责任编辑：张昕焱
封面设计：王志峰
责任校对：赵姗姗

书　　号：978-7-5611-9110-1
定　　价：228.00元

发　行：0411-84708842
传　真：0411-84701466
E-mail：dutp@dutp.cn
URL：http://www.dutp.cn